25-00.5

The BODY ATLAS

New Edition

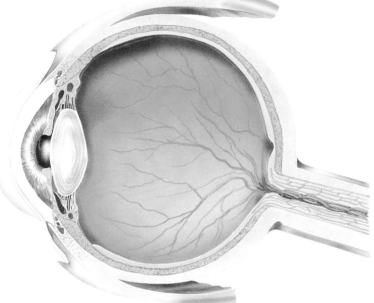

The BODY ATLAS

New Edition

Mark Crocker

Oxford

Designed by Designers and Partners, Oxford
Illustrated by Mike Saunders (Jillian Burgess Artists),
John Downes and Steve Weston
Edited by Nicola Barber

Published in the United States of America by
Oxford University Press, Inc.
200 Madison Avenue
New York, N.Y. 10016

Copublished by arrangement with Award Publications Limited/
Kim Hup Lee Printing Co. Pte Limited

Oxford is a registered trademark of
Oxford University Press

Library of Congress
Cataloging-in-Publication Data
  Crocker, Mark.
    Body atlas / Mark Crocker ; illustrated by Mike Saunders . . .
    [et al.] ; edited by Nicola Barber. —[New ed.]
      p. cm.
    Includes index.
    ISBN 0-19-520963-X
    1. Human physiology—Atlases—Juvenile literature.
    I. Saunders, Mike.
    II. Barber, Nicola. III. Title.
    QP37.C88 1992   92-11313
    612—dc20   CIP
                                            AC

Printing (last digit): 9 8 7 6 5 4 3 2 1
Printed by Cronion S.A. — Barcelona — Spain

CONTENTS

THE BODY MAP

HOW well do you know your body? Do you know what goes on inside the visible outermost layer of your skin? This *Atlas* will help you to find out more about your physical self by taking you on a tour of the body. On your journey you will learn about many of the unusual things that are going on inside your body, whether you are awake or asleep, running or sitting down.

The *Atlas* is divided into sections, each telling you about a different area of the body. Turn over to pages 8 and 9 and you will see that your body is really many different bodies, all working as one. Continue to turn the pages and you will find out how all these different parts of the body work together to make a moving, sensitive, thinking, and intelligent human being!

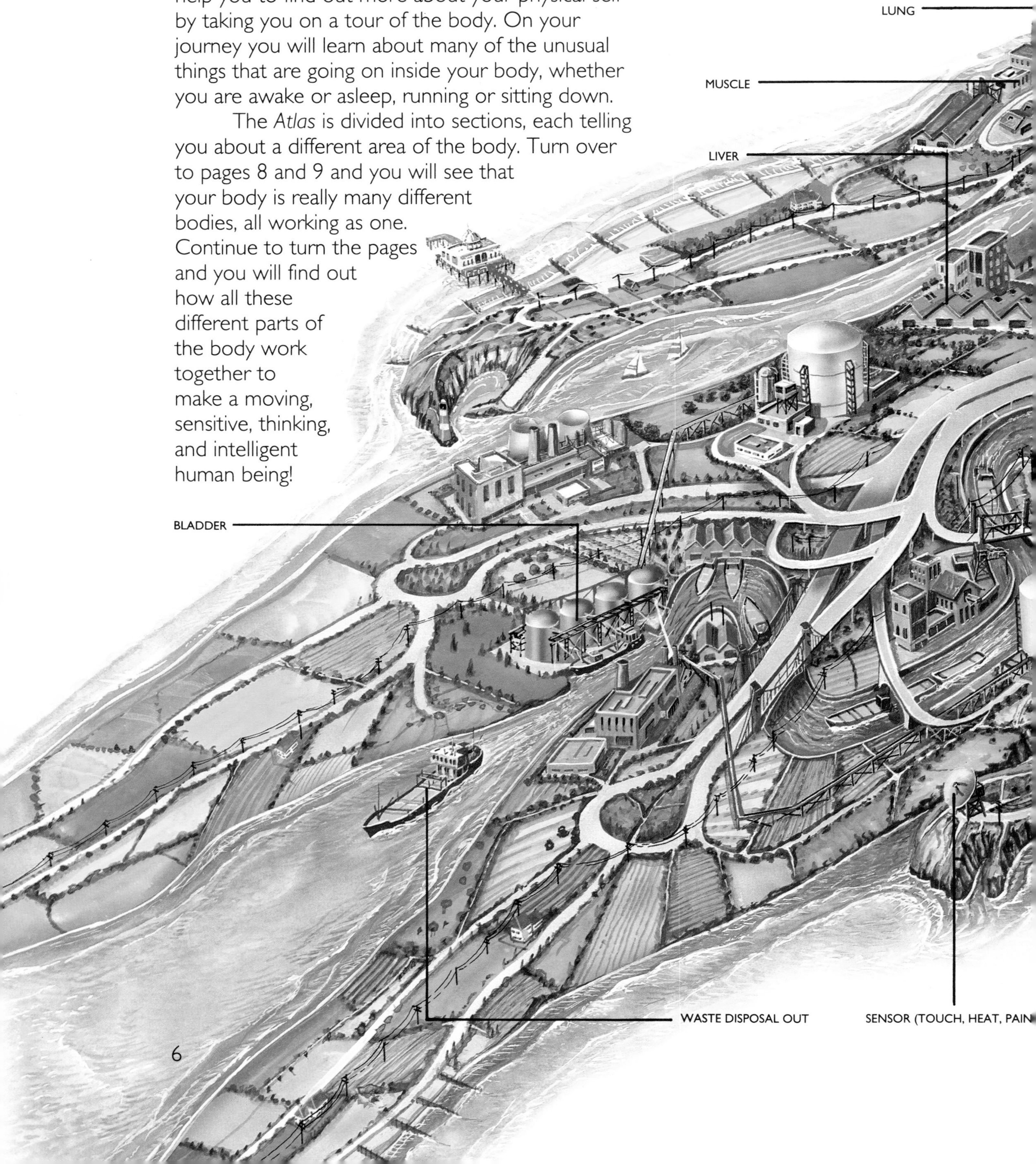

LUNG ——————

MUSCLE ——————

LIVER ——————

BLADDER ——————

—— WASTE DISPOSAL OUT   SENSOR (TOUCH, HEAT, PAIN

MOUTH

NERVES

BRAIN

BLOOD VESSELS

HEART

MUSCLE

DIGESTIVE SYSTEM

KIDNEY

The map on these pages shows how the body can be illustrated as a well-organized country. Each part of the body has a certain job. Transport around the body is the responsibility of the blood, just as road and rail systems move people and supplies around a country. In the blood are the body's police force and army — white cells and antibodies which fight off invaders and destroy any cells which are not working properly. Food is imported through your mouth to give your body energy, just as coal and oil are used to supply power stations and factories. The food is transported down your digestive system and turned into fuel, ready to be taken by the blood transport system to the centers of industry, your muscles, where the energy is used up.

The removal of waste from all this activity is the responsibility of the liver and kidneys, which act as the body's pollution controllers. For speedy communication your nerves serve as a telephone network, flashing electrical messages around the body. At the center of this well-run country is the brain, the central government of the body which thinks, makes decisions, and controls the rest of the body.

7

THE body can be divided into different "systems," each with its own part to play. Although they are illustrated separately on these pages, each system works closely with the others to maintain a healthy balance inside the body. The nervous system carries information from inside and outside the body, while the brain acts upon this information to decide what your body must do. The nerves also control the muscles which are attached to the bones of the skeleton. Your skeleton provides a framework for your body and is jointed, so that you can bend and move around.

The heart and blood together are often known as the transport system, because they move vital chemicals rapidly and efficiently around the body. The heart is the pump at the center of this system. As a backup to your blood system, the lymph system recycles any fluid that may be lost from the blood.

Your digestive system takes the food you eat and turns it into soluble substances that blood can pass on to supply the rest of the body.

The endocrine system controls slower processes in the body, such as growth and sexual changes. Your respiratory system – the lungs – takes in the oxygen that all your body needs to convert the food that you eat into useful energy.

All these systems work together to keep you alive. The reproductive systems are present to make copies of this body plan, and to produce future generations.

1 Skin *find out more about the skin on pages 44–5*

2 Nervous system *find out more about the brain and the senses on pages 34–47*

3 Transport system *find more about the heart and blood on pages 20–25*

BRAIN

SPINAL CORD

HEART

① ② ③

VEINS

ARTERIES

LYMPH VESSELS

4 Lymphatic sy
*more about the ?
system on pages*

8

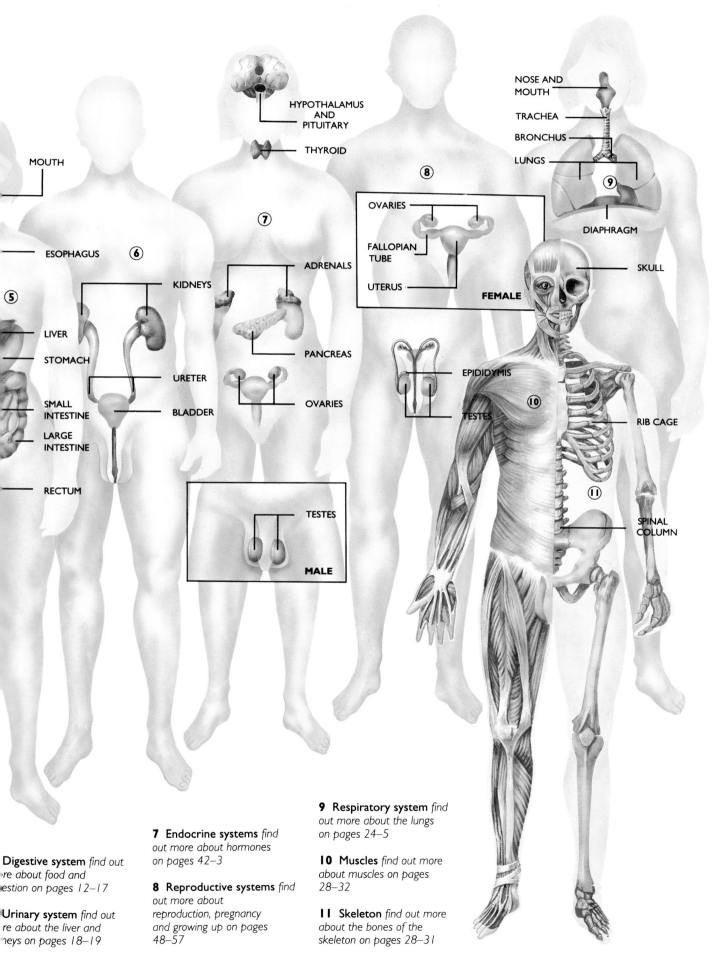

HYPOTHALAMUS AND PITUITARY

THYROID

MOUTH

ESOPHAGUS

⑥

KIDNEYS

ADRENALS

⑦

NOSE AND MOUTH

TRACHEA

BRONCHUS

LUNGS

⑧

⑨

DIAPHRAGM

OVARIES

FALLOPIAN TUBE

UTERUS

FEMALE

SKULL

⑤

LIVER

STOMACH

SMALL INTESTINE

LARGE INTESTINE

RECTUM

PANCREAS

URETER

BLADDER

OVARIES

EPIDIDYMIS

TESTES

⑩

RIB CAGE

⑪

TESTES

MALE

SPINAL COLUMN

9 Respiratory system *find out more about the lungs on pages 24–5*

7 Endocrine systems *find out more about hormones on pages 42–3*

10 Muscles *find out more about muscles on pages 28–32*

Digestive system *find out ...re about food and ...estion on pages 12–17*

8 Reproductive systems *find out more about reproduction, pregnancy and growing up on pages 48–57*

11 Skeleton *find out more about the bones of the skeleton on pages 28–31*

Urinary system *find out ...re about the liver and ...neys on pages 18–19*

9

WHEN you look at a part of your body such as the skin on your arm, it appears to be a solid piece of skin tissue. However, if you could magnify the skin enough you would see that it is really made up of millions of tiny blocks, called cells, much as a large wall is made up of many small bricks. Each cell in your body is narrower than the finest hair.

The various parts of the body all have different sorts of cells with different jobs. Skin cells, and other layers of cells, are shaped rather like bricks; muscle and nerve cells are very long and thin. Some nerve cells can be more than a yard (one meter) long but are only four ten-thousandths of an inch (one-hundredth of a millimeter) wide. The red cells in the blood are shaped like dishes and float around freely. Liver cells are large and rounded, and bone cells look like stringy spiders lying in spaces inside the hard bone. This variety of cell produces the variety of organs and tissues in the body. The way the cells are shaped, and the chemicals they produce, make you what you are.

Inside a cell

ALTHOUGH there are many different types of cells they all have a similar organization inside. You can imagine a cell as a walled town, with its own town center for control, power stations to provide energy, factories manufacturing whatever product the town specializes in, and gates in the surrounding wall that allow fuel in, and waste and the manufactured products out.

The central control of the cell comes from its nucleus. The nucleus sends out instructions to the factories (ribosomes) which

CELL V

NUCLEUS

MITOCHONDRIA

RIBOS(

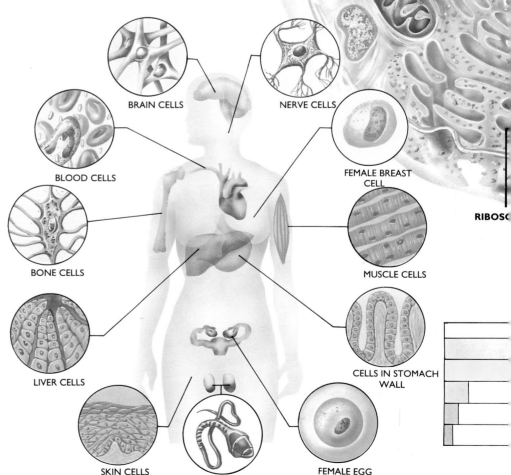

BRAIN CELLS

NERVE CELLS

BLOOD CELLS

FEMALE BREAST CELL

BONE CELLS

MUSCLE CELLS

LIVER CELLS

CELLS IN STOMACH WALL

SKIN CELLS

SPERM

FEMALE EGG

A variety of cells

The cells in each organ and system of the body differ, according to the job they have to perform. Although all the cells in the body (except red blood cells) are alike in having a nucleus, in many other respects they are very dissimilar. Nerve cells, for example, are long and thin in order to carry messages around the body quickly and efficiently. Certain specialized cells in the female breast make milk. The cells in the stomach wall produce their own mucus layer to protect themselves against acid in the stomach. (*You can see some of the other types of cell in the diagram.*)

manufacture chemicals called proteins as well as cell products which are exported out of the cell and around the body in the bloodstream. Fuel for the cell, in the form of sugars, is also carried in the blood and allowed into the cell through its wall.

The power stations in the cell are the mitochondria. They use oxygen to burn up fuel and make parcels of a substance called ATP. The ATP moves around the cell to wherever energy is needed.

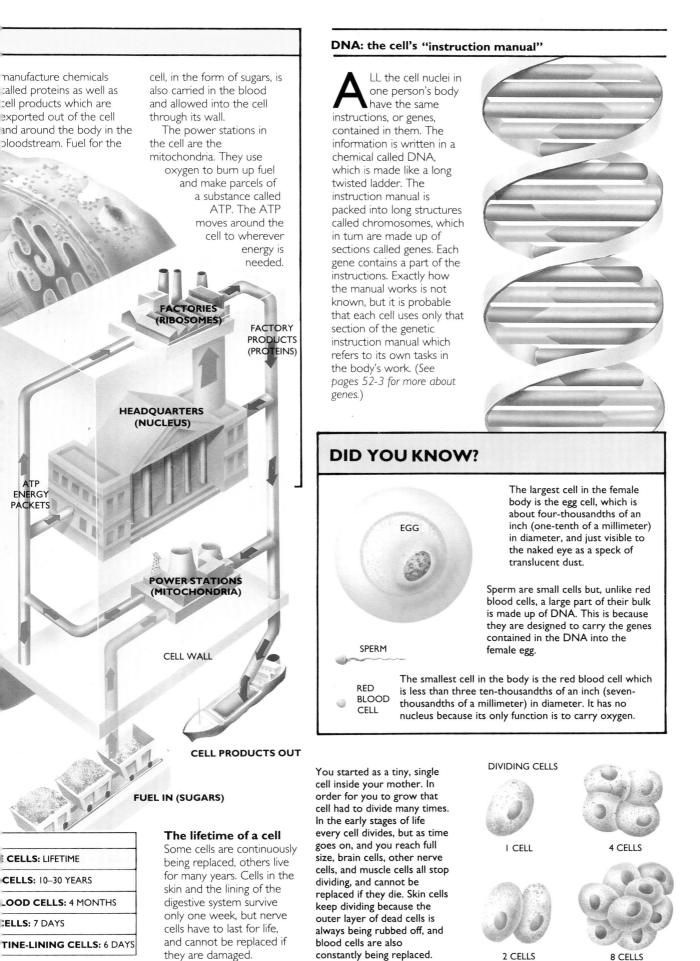

FACTORIES (RIBOSOMES)

FACTORY PRODUCTS (PROTEINS)

HEADQUARTERS (NUCLEUS)

ATP ENERGY PACKETS

POWER STATIONS (MITOCHONDRIA)

CELL WALL

CELL PRODUCTS OUT

FUEL IN (SUGARS)

CELLS: LIFETIME
CELLS: 10–30 YEARS
LOOD CELLS: 4 MONTHS
ELLS: 7 DAYS
TINE-LINING CELLS: 6 DAYS

The lifetime of a cell
Some cells are continuously being replaced, others live for many years. Cells in the skin and the lining of the digestive system survive only one week, but nerve cells have to last for life, and cannot be replaced if they are damaged.

DNA: the cell's "instruction manual"

ALL the cell nuclei in one person's body have the same instructions, or genes, contained in them. The information is written in a chemical called DNA, which is made like a long twisted ladder. The instruction manual is packed into long structures called chromosomes, which in turn are made up of sections called genes. Each gene contains a part of the instructions. Exactly how the manual works is not known, but it is probable that each cell uses only that section of the genetic instruction manual which refers to its own tasks in the body's work. (*See pages 52-3 for more about genes.*)

DID YOU KNOW?

EGG

The largest cell in the female body is the egg cell, which is about four-thousandths of an inch (one-tenth of a millimeter) in diameter, and just visible to the naked eye as a speck of translucent dust.

SPERM

Sperm are small cells but, unlike red blood cells, a large part of their bulk is made up of DNA. This is because they are designed to carry the genes contained in the DNA into the female egg.

RED BLOOD CELL

The smallest cell in the body is the red blood cell which is less than three ten-thousandths of an inch (seven-thousandths of a millimeter) in diameter. It has no nucleus because its only function is to carry oxygen.

You started as a tiny, single cell inside your mother. In order for you to grow that cell had to divide many times. In the early stages of life every cell divides, but as time goes on, and you reach full size, brain cells, other nerve cells, and muscle cells all stop dividing, and cannot be replaced if they die. Skin cells keep dividing because the outer layer of dead cells is always being rubbed off, and blood cells are also constantly being replaced.

DIVIDING CELLS

I CELL

4 CELLS

2 CELLS

8 CELLS

DIGESTIVE SYSTEM

YOU need food to keep you alive. The food you eat provides fuel for all the body's activities: every movement of the muscles as well as the continual renewal of cells that have worn out, and the parts of cells that have gone wrong.

But before the food is of any use to you, it must be broken down into pieces small enough to be dissolved into the blood. It can then be transported by the blood to the rest of the cells in the body. This process – from the moment the food enters the mouth until it crosses into the bloodstream – is called digestion. Digestion usually takes up to 18 hours, in which time the food travels down nearly nine yards (eight meters) of pipes and tubes.

MOUTH

EPIGLOTTIS

SALIVARY GLANDS

ESOPHAGUS

Where does the food go?
What happens after the food is swallowed? Inside your body are almost nine yards (eight meters) of tubes and chambers which are used to process your food. The intestines make up most of this length, coiled neatly below the stomach. It is here that most of the food is dissolved, and taken into the blood through the intestine walls.

LIVER

GALL BLADDER

LARGE INTESTINE

STOMACH

SMALL INTESTINE

RECTUM

Your digestive conveyor belt

THE digestive system is rather like a conveyor belt on a production line. After the teeth have chopped up the food in the mouth, the esophagus transports it to the stomach, where it is sterilized with acid.

The food is then treated with the first of a series of chemicals called digestive enzymes, which continue to break it up into smaller parts. For example, the white, insoluble substance in potatoes (starch) is attacked by the enzymes and broken down into sugars, which dissolve easily. The teeth and chemical enzymes are like the "crusher" in the diagram, breaking up the food into pieces small enough to go onto the conveyor belt.

In the intestines the food is on the conveyor belt itself. Here, all the useful parts of the food are gradually dissolved away through the intestine walls, like the smaller "balls" in the diagram. The larger "balls" represent the indigestible waste, which is ejected from the body through the rectum.

HOOKWORM

TAPEWORM

Your digestive system is a favorite place for parasites (animals that live and feed off other animals). If hookworms and tapeworms get into the intestines, they attach themselves with suckers, and enjoy the abundant warmth and food.

OD FROM E MOUTH

WASTE MATTER

DISSOLVED FOOD INTO THE BLOODSTREAM

TEETH AND CHEMICAL ENZYMES "CRUSH" THE FOOD

INTESTINE WALL

VILLI

Tiny fingers in the intestines

Inside the intestines are millions of tiny, fingerlike structures called villi. These increase the area that can be used to absorb the dissolved food. The cells in the villi have especially thin walls to allow the dissolved food to pass through easily. It then goes into tiny blood vessels that take the blood to a vein connecting to the liver, and on to the rest of the body. (*See page 18 for more about the liver.*)

OD VESSELS

Biting, chewing, and swallowing

TEETH are designed to bite pieces off food and to chew them into a mushy ball that can be swallowed. The front teeth cut and tear; the tougher back teeth grind and mash. The salivary glands above and below the tongue pour saliva on the food, which makes the food wet and easier to chew.

A chemical enzyme in the saliva also helps to break down starchy foods, such as bread and potatoes, into more easily digestible sugars. You can test this for yourself by chewing a piece of bread for a long time, until it tastes sweet.

The food is swallowed and travels down a tube called the esophagus. However, next to the esophagus lies the passage to the lungs (trachea), and if your food went down the wrong way, you would choke. So, a flap called the epiglottis moves across, and the food passes safely down to the stomach.

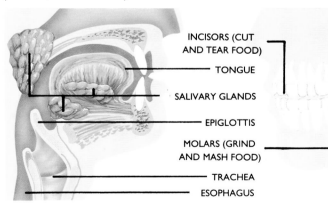

INCISORS (CUT AND TEAR FOOD)

TONGUE

SALIVARY GLANDS

EPIGLOTTIS

MOLARS (GRIND AND MASH FOOD)

TRACHEA

ESOPHAGUS

Grass and leaves are fibrous and difficult to digest, so cows eat their food twice, bringing it back from the stomach to give it a second chew. This is called "chewing the cud."

When the chewed-up ball of mushy food enters the stomach, it is treated with acid and chemical enzymes. This destroys most of the harmful bacteria, and the food begins to fall apart.

The stomach is rather like a mixer. It squashes and breaks down the food until it is soft and gooey, ready to be pushed into the long tube of the intestines. In the small intestine, new enzymes are poured in from the pancreas. By the time the food reaches the end of the small intestine it has been bombarded by 17 different enzymes.

At the end of the small intestine, digestion is complete. All the goodness has been absorbed, and the waste products are channeled into the large intestine and the rectum.

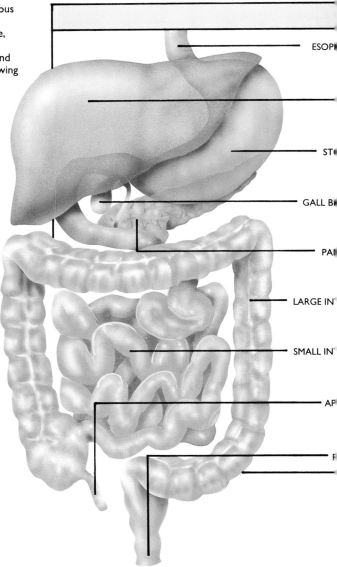

ESOP

ST

GALL B

PAI

LARGE IN

SMALL IN

AP

F

VILLI

DISSOLVED FOOD

ENZYME

FOOD

BLOOD VESSELS

SMALL INTE

Enzymes — chemicals with tiny teeth

THE most important elements in your digestive system are the enzymes which work on the food, breaking it down and turning it into substances that will dissolve into the bloodstream. Digestive enzymes attack the food, rather like little chemical teeth, biting off smaller pieces from the larger lumps of the food. These float off in the watery digestive juices to be absorbed through the thin walls of the villi, and into tiny blood vessels.

Seventeen different types of enzymes attack various parts of the food. Ten of these 17 enzymes break down protein, six work on carbohydrates, and one helps to reduce fat to tiny droplets.

OOD travels from
your mouth, down the
esophagus and into
stomach. In the small
stine, the dissolved
ducts of digestion, such
ugars and broken-up
eins, are passed
ugh the walls of the
stine into the blood.
Vaste products —
nly plant fiber and
nless bacteria — are
nnelled into the large
stine. Here, water is
ually absorbed away
these leftovers until
become solid. They
eventually ejected from
rectum.
Vhere the small and
intestines join lies a
I chamber called the
ndix. This is a leftover
the times when
ans ate only vegetation,
used it as an extra
stive chamber. It is
ss now, and can
ome inflamed if food
stuck in it, causing
ndicitis.

Gravity-free digestion

Thanks to peristalsis, food
can move through the
coiled intestines of the
digestive system even if you
are standing on your head,
or floating around in space!

As the food is swallowed, it does not simply "fall" down the esophagus — it is pushed. In fact, the mushy food is forced all the way around the digestive system by muscular movements. This process is called peristalsis, and it is going on the whole time without your being aware of it, because the muscles are controlled by nerves that work automatically. These muscular movements, as well as moving the food along the digestive tubes, also churn it around and help the process of breaking the food down.

FOOD IS PUSHED ALONG

MUSCULAR WALL   WALL PINCHES INWARDS

Pushing the food along

Food is moved along the
digestive system as the
muscles in the walls pinch
inwards. This contraction
moves along in a wave,
pushing the food in front of
it. This movement is known
as peristalsis.

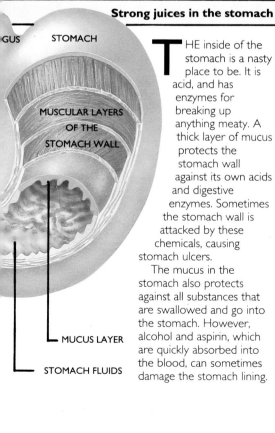

GUS   STOMACH

MUSCULAR LAYERS
OF THE
STOMACH WALL

MUCUS LAYER

STOMACH FLUIDS

Strong juices in the stomach

THE inside of the
stomach is a nasty
place to be. It is
acid, and has
enzymes for
breaking up
anything meaty. A
thick layer of mucus
protects the
stomach wall
against its own acids
and digestive
enzymes. Sometimes
the stomach wall is
attacked by these
chemicals, causing
stomach ulcers.

The mucus in the
stomach also protects
against all substances that
are swallowed and go into
the stomach. However,
alcohol and aspirin, which
are quickly absorbed into
the blood, can sometimes
damage the stomach lining.

How long does the food take to get through?

WHEN you eat
food, you do
not feel the
benefit from it immediately.
Even sugary foods, which
dissolve easily, can take half
an hour to pass into the
blood. It takes three to six
hours for a proper meal to
be broken down in the
stomach from a lumpy
mess to a slush. The first
food reaches the intestines
about an hour later.

Some of the blood that
normally goes to the brain
and muscles is diverted to
pick up the digested food.
This is why you sometimes
feel sleepy after eating. If
you exercise too hard after
eating you may get a
cramp, because there is not
enough blood in your
muscles.

The final trip through the
intestines can take between
eight and 24 hours — or
even more.

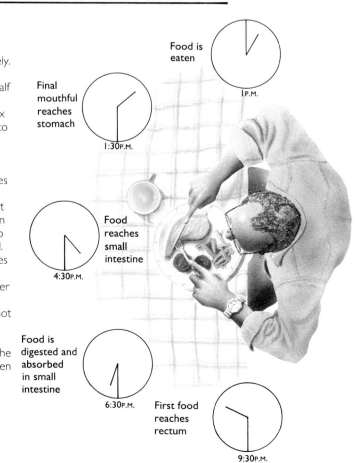

Food is
eaten

1P.M.

Final
mouthful
reaches
stomach

1:30P.M.

Food
reaches
small
intestine

4:30P.M.

Food is
digested and
absorbed
in small
intestine

6:30P.M.   First food
reaches
rectum

9:30P.M.

The sunshine vitamin

Sunshine on your skin makes a useful substance called vitamin D. Paler skinned people make vitamin D more easily than darker skinned people. Vitamin D is also obtained from foods such as butter, cheese, eggs, fish, and liver.

You need food for energy to keep your muscles working, and to keep all parts of your body in good order. Some foods (for example "junk" foods, such as candy and potato chips) have mainly energy rich substances in them (fats and carbohydrates). The energy from these foods is used up as you exercise your muscles. However, you would suffer from a disease called malnutrition if you did not also eat foods containing proteins, minerals, and vitamins, which are needed to keep the body running smoothly.

Food: the complete treatment for your body

Gasoline = Fats, sugars, and starches which give your body energy

YOUR body is like a motor car: it needs a source of energy, protective treatments, and new parts. Energy rich foods, such as fats, sugars, and starches, keep your body going — just like the gasoline in a car. The amount of energy contained in food is measured in calories. The more calories you consume the more energy you have; just as a car will go farther on a full tank of fuel.

But filling up the tank is only one part of keeping a car running. A car won't work if broken parts are not replaced. In your body proteins are used for this sort of replacement — and for growing, which is something a car cannot do. Oil changes and rust treatments also keep a car healthy. Vitamins and minerals are necessary to keep you feeling healthy, and to help the cells in your body to work properly.

Oil and rust treatments = Vitamins and minerals which keep you healthy

New parts = Proteins which also help your body to grow

Where do the calories go?

There is a healthy weight range for everyone. People who overeat and do not exercise tend to be overweight. Those who exercise strenuously and eat little are often underweight. Overeating causes food to be converted into fat, and leads to an increase in weight.

On the average you need between 1000 and 3000 calories per day, but the larger and more active you are, the more food you need.

RECOMMENDED DAILY DIET (CALORIES)	
GIRLS 12–18 YEARS	2400
WOMEN 18–35 YEARS	2100
WOMEN 35–55 YEARS	1900
WOMEN 55–75 YEARS	1600
BOYS 12–18 YEARS	3200
MEN 18–35 YEARS	2900
MEN 35–55 YEARS	2600
MEN 55–75 YEARS	2200

Eating to live or living to eat?

The risk of disease increases for both overweight and underweight people. However, in Western society more people overeat than undereat. As well as getting the energy balance right, your diet should also include enough of the essential proteins, minerals, and vitamins. The best way of achieving this is to eat a variety of foods — a balanced diet.

FOOD

MOST foods contain a mixture of carbohydrates, fat, proteins, minerals, and vitamins. You can be certain of getting the right mix if you eat a balanced and varied diet. Some foods are mainly of one type; for example, sugar is pure carbohydrate, and butter is almost all fat. The food which has the mix nearest to the ideal combination is thought to be milk. Vitamins are chemicals that are needed for a healthy body, but which you cannot make yourself, so you take them in through your food.

Proteins
Proteins are the building and replacement materials of your body, especially useful when you are growing.

Fiber
Fiber is the indigestible part of plant cells. It is needed to make the bulk of the feces, and prevent constipation.

Fats
Fatty and oily foods are the highest in calories, so you need to be careful not to eat too many.

Minerals
Vital minerals include calcium for bones and teeth, and iron for red blood cells.

Carbohydrates
Like fats, carbohydrates are high energy foods. They include sugars and starches, and make up the bulk of most plant foods.

Vitamin A
Used in eye pigment and the lungs, it is rarely in short supply.

Vitamin B
Vitamins in this group affect cell division, and release energy from food.

Vitamin C
Needed in skin and joints for healing wounds. Deficiency causes bleeding of gums and joints (scurvy).

Vitamin D
Used in the formation of bones. Lack of it causes soft, deformed bones (rickets).

Vitamin E
Thought to be necessary for healthy reproduction.

Vitamin K
Used in blood clotting. It is made by bacteria in the intestines as well as being supplied by food.

The liver is a large organ, lying just under the lungs. It acts as a general cleaner and regulator of the blood: your pollution controller. It ensures that poisons are made harmless, takes out dissolved foods and removes dead cells. In this way, all the garbage in the blood is either disposed of, or recycled.

The liver also stores foods, such as sugars, and releases them as necessary, especially during exercise. If you overeat, the liver sends some of the food to be laid down as fat, and the rest it converts to waste chemicals which go to the kidneys.

Alcohol is a poison which is converted to less harmful chemicals by the liver. Small amounts can be dealt with easily, but too much alcohol, taken too often, can cause liver damage.

THE LIVER

LIVER

Sugar in your blood
You need the right amount of sugar in your blood because all the cells in the body use sugar for energy. Your sugar levels are controlled by the liver, and by a chemical called insulin.

Diabetes is a disease that causes this control to fail. Sometimes, people with diabetes have sudden drops in their sugar levels, which can make them very ill. However, regular injections of insulin help to balance the blood sugar levels, and usually control diabetes.

THE liver is the only organ which has a vein running to it that does not go directly to the heart. This vein carries blood with food in it from the intestines. The liver takes in the dissolved food, extracts the chemical waste and stores the useful food and vitamins to be used as the body needs them.

Some of the chemical waste is turned into bile, which is stored in the gall bladder and released into

GALL BLADDER

LIVER

BILE DUCT

PANCREAS

Portal vein carries blood with food in it from the intestines

ARTERY

Stones in the kidneys: a simple solution

THE body needs sufficient liquid to make the kidneys work properly. Sometimes, if there is a lack of fluid, the kidneys make urine which is too concentrated, and kidney "stones" form from the solids in the urine. These block the ureters and cause extreme pain unless they are removed.

Most kidney stones can now be removed without an operation. By using a machine that makes high-pitched sound waves, the stones are gradually broken down into sandlike particles which are passed out with the urine.

Kidney stones come in all shapes and sizes

e intestines. Bile not only
removes waste, but also
helps with digestion by
making fats soluble.

About one quart (one
liter) of blood is pumped
through the kidneys every
minute. It is forced into tiny
vessels in the kidneys and
filtered. The purified blood
passes back to the heart;
the waste, or urine, trickles
down the ureters to the
bladder, which blows up
like a balloon until it is full.

People with diseased or
damaged kidneys will
eventually die if their blood is
not purified regularly. A
kidney machine can
substitute for their kidneys.
Tubes are joined to their
blood vessels, and the blood
passes through the machine
for filtration.

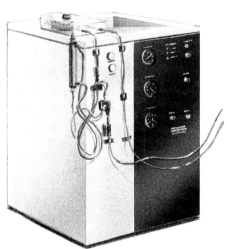

DNEY

EIN

RTERY

O BLADDER

URETER

Your blood is like the water in a river — it must be
kept clean and free of poisons. The kidneys are
your garbage collectors, filtering anything that
is not wanted from the blood, so that it can
pass out of the body as urine. The kidneys
get rid of all the bits you do not need,
retaining only the useful substances
for the blood. They also control the
levels of salts and water in the
blood. Without this filtering
process, your blood would
become polluted and kill
you, just as dirty water in a
river can poison its plant
and animal life.

KIDNEYS

URETER

BLADDER

THE KIDNEYS

Surviving in the desert
Water is precious if you
live in the desert. This
means that desert animals
must keep as much water
as possible inside their
bodies. Animals such as the
kangaroo rat do this by
increasing the amount of
salt in their urine, and
decreasing the amount of
water lost.

DID YOU KNOW?

As much as 90 percent of the
liver can be destroyed before
it is damaged permanently.
The remaining ten percent
can regrow back to a
complete liver of the normal
size.

Some people excrete a
strong-smelling substance in
their urine after eating
asparagus, because of the
way in which some of the
chemicals present in the
asparagus break down in the
stomach.

The liver stores
vitamins such as A and
D in large quantities.
You can have too
much of a good thing
though — a polar
bear's liver stores
enough vitamin A to
kill a human.

Urea makes urine yellow and
is produced in the liver as
proteins are broken down. If
you eat a very meaty meal
(high in protein), your urine
is likely to show it later by
being dark yellow.

TRANSPORT SYSTEM

THE body can be compared to a large country made up of millions of tiny walled villages, or "cells." All these cells must be supplied with everything they need to keep functioning: sugar for fuel, oxygen to burn up the fuel, and proteins for building and repair work.

The blood is your transport system. Pumped by the heart, it carries these essential supplies all over the body. Its most important deliveries are oxygen and dissolved food. Oxygen is collected from the lungs and carried in the blood in red cells. Dissolved food is taken from the intestines to the liver, and from there distributed around the body as required.

Why do you feel slee
after eating?
Sometimes there may be problems transporting enough blood to all parts of the body, especially if too many demands are made at once. After a meal, extra blood flows the digestive system to pick up food. This means there is less blood and le oxygen for your brain ar your muscles, so you ma feel sleepy.

Tiny blood vessels
The "main roads" of your blood transport system are the veins and arteries, but to reach the most remote cells the blood must travel along tiny blood vessels called capillaries. It has been estimated that the blood vessels and capillaries from one person, laid end to end, would stretch around the Earth twice.

CAPILLARIE

PLASMA

WHITE CELLS

CLOTTING
PARTICLES

RED CELLS

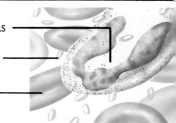

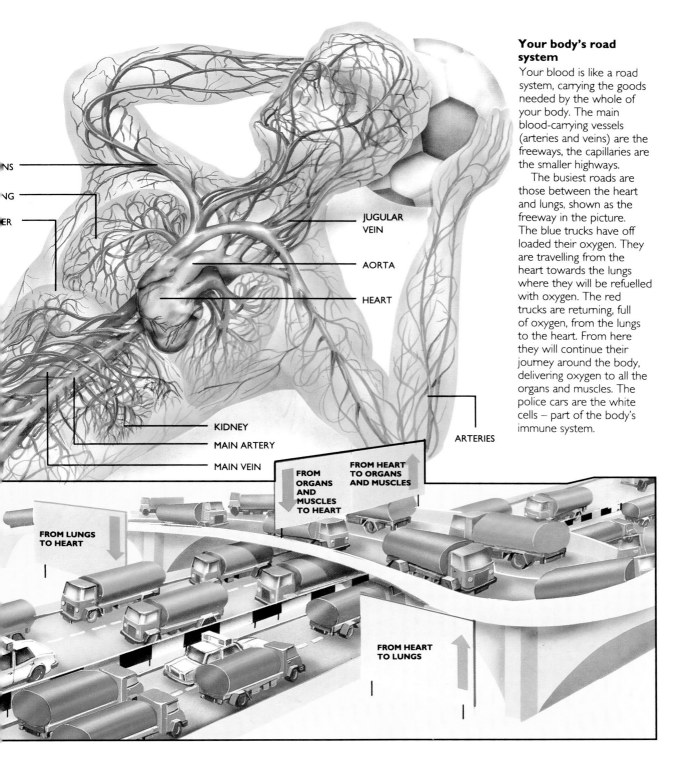

NS ——

NG ——

ER ——

JUGULAR
VEIN

AORTA

HEART

KIDNEY

MAIN ARTERY

MAIN VEIN

ARTERIES

FROM
ORGANS
AND
MUSCLES
TO HEART

FROM HEART
TO ORGANS
AND MUSCLES

FROM LUNGS
TO HEART

FROM HEART
TO LUNGS

Your body's road system

Your blood is like a road system, carrying the goods needed by the whole of your body. The main blood-carrying vessels (arteries and veins) are the freeways, the capillaries are the smaller highways.

The busiest roads are those between the heart and lungs, shown as the freeway in the picture. The blue trucks have off loaded their oxygen. They are travelling from the heart towards the lungs where they will be refuelled with oxygen. The red trucks are returning, full of oxygen, from the lungs to the heart. From here they will continue their journey around the body, delivering oxygen to all the organs and muscles. The police cars are the white cells – part of the body's immune system.

Inside your bloodstream

Although it looks liquid, your blood is actually made of millions of tiny cells floating around in a watery

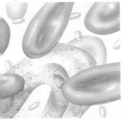

substance called plasma. If a sample of blood is left to settle for a few hours in a test tube, these cells appear as a solid lump at the bottom.

Most of the blood cells are red cells which are used for carrying oxygen and live only a few weeks. The remaining cells are the white cells, which defend you against invading germs.

Your almost five quarts (five liters) of blood are made up of many different parts. Much of the blood consists of red oxygen-carrying cells which give the blood its red color. There are also white cells which work to protect you against invading germs such as bacteria and viruses (see pages 26-7). The blood even has a system of garbage collection, carrying away waste chemicals from the cells to the liver and kidneys (see pages 18-19).

21

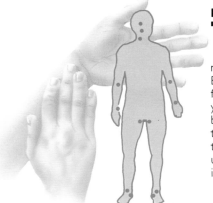

THERE are places around your body where big arteries lie near the surface of the skin. By touching them, you can feel the pumping action of your heart pushing the blood along. This is called the pulse. The easiest places to feel it are on the underside of your wrist or in your neck.

The heart is at the center of the blood transport system. It is a powerful machine, pumping blood all around the body. With 50,000 miles (80,000 kilometers) of capillaries to push the blood through, this amazing organ needs to be tough. It is made of a thick, strong muscle which moves automatically, and can work over 100 years without a break! Controlled by the brain, your heart beats at different rates depending upon how active you are and how much oxygen your muscles need. When you are resting, your heart beats between 60 and 70 times a minute. When you are active – running or swimming – your heart beats faster in order to pump more blood around your body.

VEINS: TO HEART

VEIN

ARTERIES: FROM HEART

ARTERY

AR

(A)

(C)

AORTIC VALVE

BICUSPID

TRICUSPID

(B)

(D)

PULMONARY ARTERY

PULMO

LIVER

LUNGS

STOMACH AND INTESTINES

VEIN

KIDNEYS, GENITALS AND LEGS

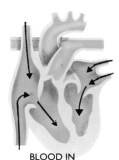

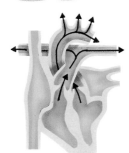

BLOOD IN

BLOOD OUT

1. The heart muscles relax and blood is sucked in through the pulmonary vein from the lungs, and through the veins from the rest of the body.

2. When it is full of blood, the heart contracts and as a result blood is pumped out along the pulmonary artery to the lungs and along the other arteries to the rest of the body.

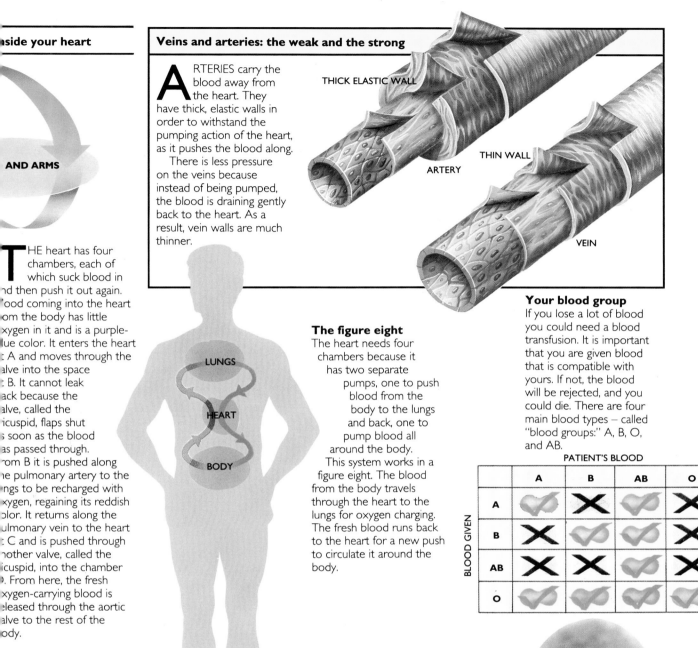

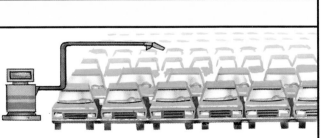

AND ARMS

THE heart has four chambers, each of which suck blood in nd then push it out again. lood coming into the heart om the body has little xygen in it and is a purple-lue color. It enters the heart t A and moves through the alve into the space : B. It cannot leak ack because the alve, called the icuspid, flaps shut s soon as the blood as passed through. rom B it is pushed along ie pulmonary artery to the ings to be recharged with xygen, regaining its reddish olor. It returns along the ulmonary vein to the heart t C and is pushed through iother valve, called the icuspid, into the chamber). From here, the fresh xygen-carrying blood is eleased through the aortic alve to the rest of the ody.

Veins and arteries: the weak and the strong

ARTERIES carry the blood away from the heart. They have thick, elastic walls in order to withstand the pumping action of the heart, as it pushes the blood along.

There is less pressure on the veins because instead of being pumped, the blood is draining gently back to the heart. As a result, vein walls are much thinner.

THICK ELASTIC WALL

ARTERY

THIN WALL

VEIN

The figure eight

The heart needs four chambers because it has two separate pumps, one to push blood from the body to the lungs and back, one to pump blood all around the body.

This system works in a figure eight. The blood from the body travels through the heart to the lungs for oxygen charging. The fresh blood runs back to the heart for a new push to circulate it around the body.

LUNGS

HEART

BODY

Your blood group

If you lose a lot of blood you could need a blood transfusion. It is important that you are given blood that is compatible with yours. If not, the blood will be rejected, and you could die. There are four main blood types – called "blood groups:" A, B, O, and AB.

PATIENT'S BLOOD

	A	B	AB	O
A	✓	X	✓	X
B	X	✓	✓	X
AB	X	X	✓	X
O	✓	✓	✓	✓

BLOOD GIVEN

DID YOU KNOW?

our heart, which weighs less than ine pound (half a kilogram), will eat about three billion times and ump about 800 million pints 400 million liters) of blood during our lifetime. This equals the mount of gasoline it would take to ll 10 million cars.

ou blush when you are hot or embarrassed because the blood runs nto the capillaries on your face, illowing body heat to escape. A red ilush acts as an emotional signal to show others that you are upset.

our blood transport system is like a central heating system. It spreads heat from warm areas of the body o cooler areas.

Your heart has to work harder to push the blood up to your head than down to your legs. If you stand up too suddenly after resting for a while, your blood pressure could be too low for blood to reach your head, and you might feel faint.

Queen Victoria passed on to several of her grandchildren a disease called hemophilia: blood would not clot after injuries. Hemophilia can now be treated, but in those days, it could cause death. (See page 45 for more about how a wound heals.)

The cells in your body need oxygen to stay alive. Oxygen is used to burn up sugars, which provide energy, but this process produces a waste gas, carbon dioxide, which must be taken away again. The problem is solved by breathing in oxygen through the lungs, and breathing carbon dioxide out again.

The lungs are like a giant sponge, with millions of tiny chambers for the air to mix with fine blood capillaries. When you breathe in, the oxygen-rich air is sucked into the tiny chambers of the lungs. The oxygen passes through the thin walls of these chambers and dissolves in the blood, and carbon dioxide waste escapes. The blood is now ready to transport the oxygen from your lungs to the rest of your body.

It is most important that your brain and your muscles have a plentiful supply of oxygen, to keep you awake, and thinking, and active. You ▶

Inside the lungs of an adult male, the surface area of the alveoli is huge – spread out flat they would cover a tennis court.

This vast network means that the gases can cross over from the blood to the lungs very quickly, making respiration – the processes of breathing – very efficient.

Journey down your windpipe

YOU normally breathe in through your nose, where hairs and mucus trap the dust and dirt particles in the air, but you can also breathe through your mouth when necessary. At the top of your windpipe, called the trachea, is the epiglottis, a flap of tissue that prevents food from going the wrong way and blocking the trachea. This opens to allow the air through. The trachea splits into two bronchi, one for each lung. These split even further into smaller and smaller tubes, finally stopping in millions of tiny dead ends called alveoli. This is where

NOSE

MOUTH

BRONCHUS

BRONCHIOLES

RIGHT LUNG CUT OPEN

ALVEOLI

CAPILLARIES

ALVEO WALL

ARTE

BRONCHIOLE

VEIN

How long can you hold your breath?

MOST of the time you are not aware that your breathing is being adjusted to suit the amount of work your muscles are doing. If you hold your breath you override this normal control. The oxygen levels in your blood will drop, and carbon dioxide levels rise. After 30 seconds or so of holding a breath, most people will need to breathe again. Pearl divers can stay underwater for much longer times because they train to make their lungs and hearts more efficient.

You have about seven million alveoli, each about the size of a pinhead, in your lungs. They look like miniature bunches of grapes, and have delicate, thin walls so that the oxygen and carbon dioxide can pass through easily.

24

the waste carbon dioxide in the blood is exchanged for oxygen.

The rib cage surrounds and protects your lungs; below this is a muscular layer called the diaphragm. At the top of the trachea lies the larynx – your voice-box. As air is breathed out, it vibrates the stringy vocal cords in your larynx, producing your voice sounds.

If you get dust or pollen on the sensitive parts of your nose a powerful spasm in the breathing muscles pushes the air out at around 100 miles an hour. The particles causing the sneeze are removed.

have the potential to make scientific discoveries, play music, write books, run marathons, climb mountains, and play football and tennis all because your lungs and blood deliver the oxygen to your muscles and brain, quickly and efficiently.

Loading up the oxygen
The recharging process in the lungs is rather like trucks refuelling at a gas station. The trucks arrive empty of oxygen, which gives them a blue color. They load up with oxygen and their color changes to red.

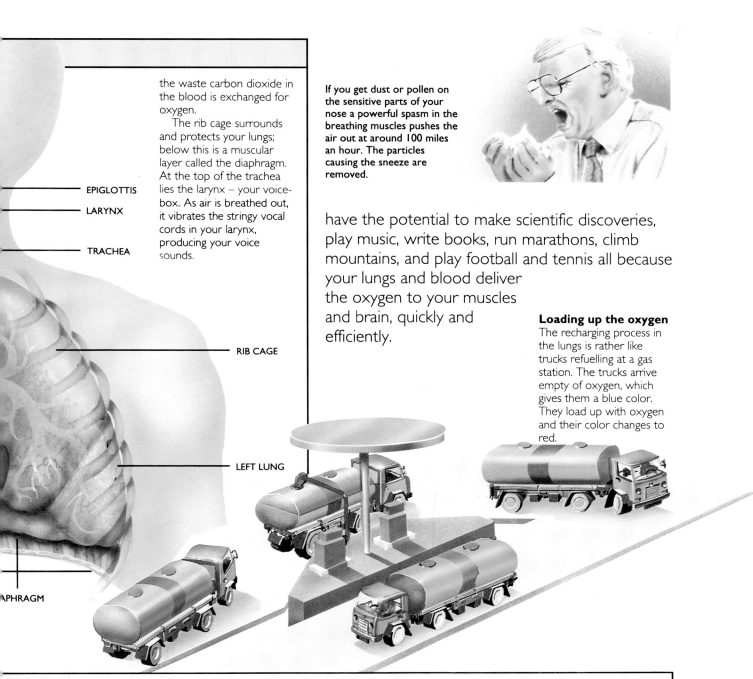

EPIGLOTTIS

LARYNX

TRACHEA

RIB CAGE

LEFT LUNG

DIAPHRAGM

Breathing in and breathing out

YOU breathe in using a combination of your rib muscles and your diaphragm. The muscles outside the ribs shorten, pulling the ribs upwards and outwards. The diaphragm also shortens, pulling down. These actions make your chest and lung cavity bigger, and air is sucked in to fill the space created.

To breathe out, the muscles relax and your elastic lungs return to their original size, emptying the air out as they do so.

Your brain automatically controls the rate and the depth of your breathing, according to the needs of your body. Normally you take in about half a quart (half a liter) of air at a time, but if your muscles are working hard, and you need more oxygen, your breathing rate increases and you can breathe in two or three quarts (two or three liters) of air at a time.

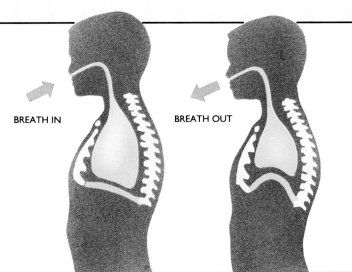

BREATH IN

BREATH OUT

The boy in the bubble

SOME unfortunate people are born with an immune system that does not work. The longest anyone has lived with this problem is 12 years. An American boy, called David, was kept alive by being enclosed in a plastic bubble shortly after birth to keep all germs out. His food and drink were sterilized, air was filtered, and no direct physical contact was allowed. The immune system is your plastic bubble protecting you from a myriad of germs.

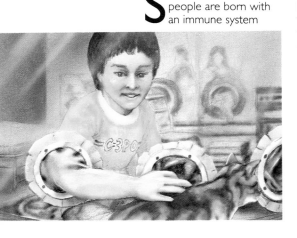

What causes an allergy?

YOUR white cells are good at recognizing invading germs, but sometimes they make mistakes and respond to harmless substances. This is what happens to people who suffer from hay fever. Their white cells mistakenly attack pollen, produced by grasses and trees in early summer, in the sensitive parts of the nose and eyes. This causes swelling and severe itching.

Dust from house mites, insect bites, stings, and pet hair can also cause allergic reactions.

HOUSE MITES

WASP STINGS

PETS

POLLEN

Your body is an ideal place for germs and other microscopic creatures to live. These invaders can feed off your body fluids and even the insides of your cells. Although you have protective barriers, such as a tough layer of cells on the skin, and acid in the stomach, many of the invaders still survive and would kill you very quickly if you did not have an army to defend you.

The white cells in the blood are your defense against infection. They can recognize invaders and send ammunition to kill them. Antibodies are the white cells' missiles that attack the germs. They also signal to the body's tanks, large white cells called macrophages, to surround and destroy the germs. Together, these make up the body's immune system.

26

Your immune response

THE first stage in fighting off an invading disease is to recognize the intruder. This is the job of one type of white cell – the T-cell. It goes around the body testing all the objects it meets and will stick to any invaders, ringing alarm bells and sending for another type of white cell, the B-cell. The B-cell makes deadly chemicals, called antibodies.

Millions of different antibodies are present inside you, each one made by a different B-cell. Each one is designed to kill only one sort of germ. The T-cell tries out thousands of B-cells until it finds the one that makes the right antibody to match the germ. When it has completed its search, it tells the B-cell to divide up into a mass of cells, all making a supply of antibodies to kill the invading germs.

Once the germs are coated in antibodies that is the signal for the final stage – a white cell called a macrophage engulfs and destroys them.

4. The antibodies produced by the B-cell coat the germs, signalling to the macrophage to engulf them.

3. The T-cell has found the right B-cell. It tells the B-cell to divide and produce antibodies.

2. The T-cells try out B-cells to find the right antibody for the invading germs.

1. Germs enter body, throug cut, in the ai lungs, or in f the intestine cells recogni: germs as inv:

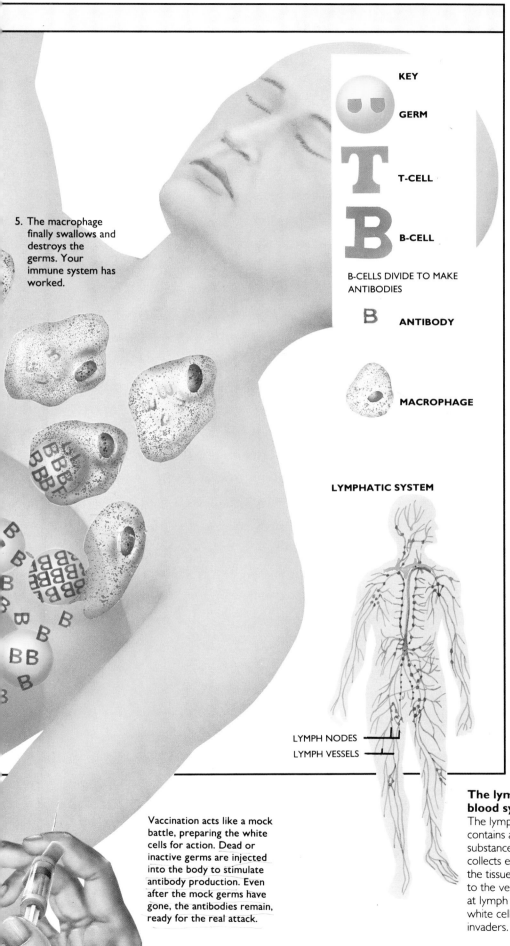

5. The macrophage finally swallows and destroys the germs. Your immune system has worked.

KEY

GERM

T T-CELL

B B-CELL

B-CELLS DIVIDE TO MAKE ANTIBODIES

B ANTIBODY

MACROPHAGE

LYMPHATIC SYSTEM

LYMPH NODES
LYMPH VESSELS

Vaccination acts like a mock battle, preparing the white cells for action. Dead or inactive germs are injected into the body to stimulate antibody production. Even after the mock germs have gone, the antibodies remain, ready for the real attack.

The discovery of vaccination

Before vaccination was discovered, diseases such as tuberculosis and smallpox, which are virtually unknown today, killed many people. The first person to realize the importance of vaccination was an Englishman, Edward Jenner. He noticed that people who worked with cows often suffered from cowpox, a disease similar to smallpox but less severe, and that they rarely caught smallpox itself. He argued that the cowpox germ gave the body resistance to smallpox.

He had so much faith in his method that, in 1796, he smeared cowpox pus onto the scratched arm of a boy called James Phipps. He was right – the cowpox "vaccination" triggered off the right antibodies to kill smallpox, and the boy remained immune all his life.

The lymph: a backup blood system

The lymph system, which contains a clear, watery substance called lymph, collects excess liquid from the tissues and returns it to the veins. It filters germs at lymph nodes where the white cells attack the invaders.

27

SKELETON AND MUSCLES

I F YOU did not have any bones you would be
soft, like jelly. The bones that make up your
skeleton give you support and protection against
injury. But if your body was only made of bones
you would be like a puppet without strings -
collapsed in a heap. Of course, you do not have
strings pulling from above, but you do have
internal strings: the muscles. The muscles are
attached to the bones and hold them in place.
Muscles also allow you to move the bones when
necessary, by a shortening of the muscle that, in
turn, moves the bone.

What are muscles?
There are muscles all over
the body, wherever the
bones can move one against
the other. Some are near
the surface, such as on the
chest where the muscles
control the shoulder and
arm. You can feel these
muscles tense and bulge as

you move your upper arm.
Some muscles are deep
inside the body and are
more difficult to find. In
between your ribs there are
muscles which help you to
breathe by pulling the ribs
together for breathing out,
and pulling them apart for
breathing in.

FEMUR

The cyclist can reach speeds of
up to 30 miles an hour
with the power from the
muscles in his legs.

28

The brain protector
The skull provides protection for the brain. The bones are joined together to form a rigid casing with a hole at the base to allow the spinal cord to pass out, carrying all the nerves down to the rest of the body. The skull also has holes to allow signals to reach the brain from the eyes, ears, and nose.

SKULL

Using your hands
In the hand, the muscles and bones are intricately connected to form a structure that is capable of many different actions. This picture shows the bones, and some of the muscles which are connected to the fingers by long tendons.

RIB CAGE

The rib cage surrounds and protects your lungs.

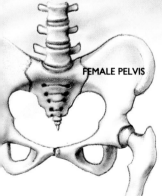
FEMALE PELVIS

The spine
The spine takes the full weight of the upper body, connecting it to the lower part. The backbone is made up of many small bones or *vertebrae* resting one on another. The bones can move over each other slightly to allow the body to bend. The spine also curves naturally to allow for this movement. However, too much curvature of the spine can cause injury to the bones and bad posture (the way you stand or sit). That is why you are told to "sit up straight."

Male or female hips?
Male and female bones differ slightly all over the body, but most of all in the pelvic (hip) region. A woman has broader hips than a man to allow babies to be born.

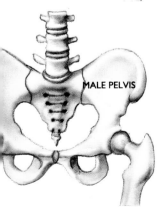
MALE PELVIS

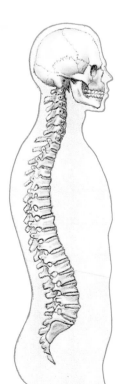

29

Joining the joints

Joints would fall apart if they did not have tough ropelike straps, called ligaments, to bind them together. If you damage any of these, the joint could pull apart or tear. The inside of the joint has a slippery liquid to lubricate the moving parts, like oil in a car. As well as this lubricant, the joints also have a layer of smooth, hard material, called cartilage, covering the ends of the bone.

KNEE JOINT
CARTILAGE
JOINT FLUID
LIGAMENT

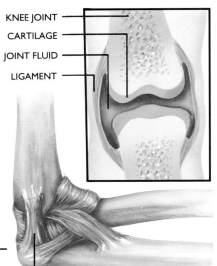

ELBOW JOINT

LIGAMENTS

Joining up your bones

YOUR bones are tough and rigid in order to give you support; but, of course, you must be able to move too. Many of the bones are joined by flexible parts, called moving joints. There are also some fused bones, such as those in the skull, where the bones are used for protection rather than movement.

Some joints move only a little, giving strength and

The two neck vertebrae at the top of the spine join wit the skull. They have a swivel joint in them which allows th neck to twist around.

The elbow and knee joints are hinge joints which allow movement up and down, rather like a door hinge.

Imagine trying to build a model robot of a person, capable of moving just like a human. Your robot must be able to write, ski, play football, perform surgery, and play the violin. No robot is that good. As you look closer at the skeleton and its joints, you will realize that each part is shaped and fitted for a specific purpose. All 208 pieces of the skeleton fit together perfectly, allowing for just the right amount of twisting, pushing, or pulling needed in just the right place. You are a marvelous construction of solid parts that gives the softer muscles and skin the support and attachments that they need.

As well as providing a framework, your bones also protect many of the delicate, soft organs such as your brain and lungs. For this reason, your bones must be strong, and they are – four times tougher than concrete!

Inside a bone

The outer layer of bone, the periosteum, a whitish skin which ha both nerves and blood it, and is sensitive to pair The next layer is the har compact bone. This is the toughest part of the bone, and has to be sawed through in operations. The compact bone is hollow an is lined with spongy bone, which combines lightness and strength.

At the center of some bones lies a jellylike substance called bone marrow. It is important because it manufactures all the blood cells in your body.

Tough teeth anchored in bone

The tooth is protected by a hard layer of dentine, and an even tougher layer of enamel, which is white, and the hardest substance in the body. A bony mixture called cementum glues the tooth into the bone in the jaw.

GUM
ENAMEL
DENTINE
CAVITY
BONE

CEMENTUM

Some bones, such as the vertebrae in the backbone and the small bones in the ankle, can slide over one another. Each bone moves only a little, but because there are many of them, they allow quite a lot of flexibility.

rigidity with a small amount of flexibility. The bones in the backbone (spine) are connected in this way. But because there are 26 of them, each allowing a little movement, you can actually twist your spine enough to bend over, or to look behind your back.

The moving joints in the body all have a special shape depending upon their job. In the elbow, there is a hinge joint which moves in one direction, up and down, rather like the hinge of a door. Plane joints in the ankle allow very little free movement, except for a small amount of sliding. Ball and socket joints, in your shoulders and hips, allow the freest movement of all.

VERTEBRAE

SPINE

Ball and socket joints are found in the hips and shoulders. They allow movement in all directions, limited only by the muscle and ligament attachments.

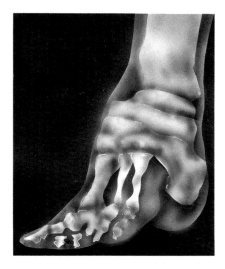

Beautifying the bones?
Damage to the body in pursuit of beauty has often been practiced. In some areas of the world it is still traditional for women to wear rings around their necks, which appear to elongate the neck. Some Chinese women used to bind up their feet to squash the bones and keep them small (left). In England, Victorian ladies had ribs removed in order to fit into corsets that made their waists look thin.

Healing a break

ALTHOUGH they are very strong, your bones can break if you put too much strain on them. Fractures, or breaks, are divided into two main types. A simple fracture (1) only involves the bone, but a compound fracture (2) happens when the bone has broken and pierced the skin.

Other more complicated fractures occur when one bone has become pushed into another (3), or when the bone shatters inside the body (4).

An X-ray, which produces a picture of the skeleton, is used to reveal broken bones. Although the bone will heal itself, it is usually necessary to put a plaster cast or splint around the affected area in order to support the bone, and to make sure it heals in the correct position.

① ② ③ ④

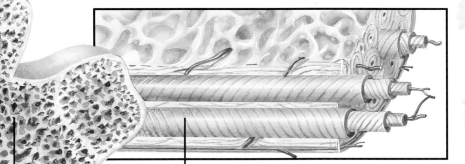

PROTEIN FIBERS

SPONGY BONE

PERIOSTEUM

BONE MARROW

COMPACT BONE

BLOOD VESSEL

Your bones are made of a hard limestonelike mixture of salts. Running through this substance are long, tough protein fibers, rather like the steel rods that run through reinforced concrete to make it stronger. Bones are formed in a cylindrical shape, which not only gives extra strength, but also makes them lighter. If a bone breaks, it will usually repair itself.

Your skeleton, with its complex joints and intricate parts, provides the framework for your body, but without muscles it would not be able to move. Muscles are unusual tissues that can convert the energy stored in food into pulling power. They are attached to the bones by ropelike tendons, and are controlled by the nerves which, in turn, receive instructions from the brain. At the right signal the long, thin cells inside the muscle become shorter and thicker, and as the muscle shortens it pulls the bone to which it is attached. Muscles can only pull — they never push. Sometimes it can take a bit of practice to make your muscles do exactly what you want them to. Riding a bicycle, playing a musical instrument, or walking a tightrope all require the coordination and training of many different muscles. There are also muscles which you ▼

BICEPS SHORTENS

Moving your arm
The muscles in your upper arm provide a perfect example of how muscles often work in pairs, one set pulling against the other. To lift a weight the larger biceps on top of the arm shortens. This lengthens and relaxes the triceps at the back. To drop the arm the triceps shortens, pulling the biceps back to its original shape.

TRICEPS LENGTHENS

BICEPS LENGTHENS

TRICEPS SHORTENS

cannot control. The tireless muscles in the heart (see pages 22-3) and intestines (see page 15) work automatically throughout your life, pumping blood and pushing food.

Looking inside a muscle

LIKE the rest of your body, your muscles are made up of cells. Your muscle cells are unusual because they are extremely long and thin. They can measure up to an eighth of an inch (two centimeters), but they are less than the width of a hair. They produce their power by shortening when stimulated by a nerve. The cells are packed together to make up fibers which are themselves combined to make the whole muscle. When the cells shorten,

the whole muscle does, too.
The actual power comes from the thin fibers inside the muscle cell. At a signal from the nerve these fibers move towards the center of the cell, rather like long, thin rowing boats using tiny oars to paddle towards each other. This makes the cell shorter and fatter, and shortens the whole muscle, pulling the bone to which it is attached.

LONG MUSCLE CELL SHORTENING
"OARS" BETWEEN FIBERS
THIN FIBERS

SHORT, FAT MUSCLE CELL

DIFFERENT TYPES OF MUSCLE CELL

Lifting weights
Weightlifters are very careful to use their powerful thigh muscles to lift, and not to bend their backs. The bones and muscles in the back can stand strain straight down, but move out of shape if they are strained sideways or forwards. Keeping a straight back is important when lifting any weights, small or large.

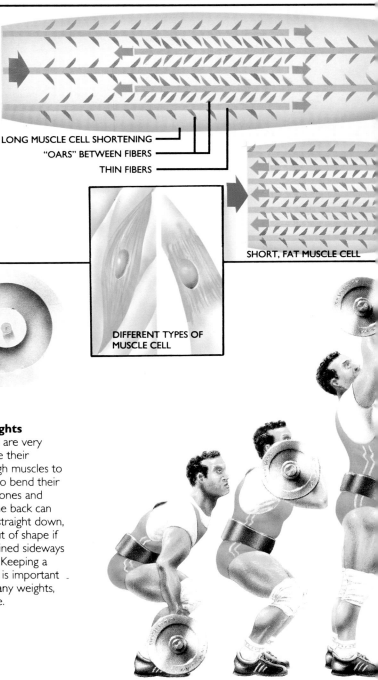

SPARE PARTS

WHAT happens if a joint wears out, or your kidneys go wrong? Today, it is possible to have a spare part replacement. Spare parts can replace bones, organs, and even whole limbs, but such operations are usually only carried out as a last resort. It is possible to replace someone's heart completely, by transplanting a heart from another person, but this is an extremely difficult and skilled operation. Even more amazing are heart and lung transplants in tiny babies, when the baby is kept cold with ice to slow down its heartbeat and allow the operation to work.

Some people have spare part replacements to improve their appearance. Skin can be grafted from one area of the body to another, noses can be remodelled, and faces can be rebuilt completely. It is even possible to transplant hair onto bald heads.

Artificial parts
Artificial legs and arms have been used for many years, but there have been many improvements since the wooden leg was invented. Modern devices are difficult to detect under clothes, and are especially effective if only part of the limb is artificial. It is even possible to produce an arm with a mechanical hand, activated by the shoulder muscles.

Elbow, hip, knee, and shoulder joints can all be replaced when they wear out. The problems caused by the body rejecting these new parts have been overcome by using plastics, and metals such as titanium. The new joint is cemented into the bone once the old joint has been removed.

A new heart
An artificial heart is often not as effective as a real one, transplanted from another human.

A transplanted heart from someone who has recently died can give fresh life and vigor to the person who receives it. But an artificial heart is often rejected by the body. Other parts of the body which can be transplanted include the kidneys, liver, skin, cornea of the eye, and hair.

Replacement parts
1. Wig
2. Skull plate
3. Eye
4. Ear
5. False teeth
6. Larynx
7. Shoulder joint
8. Lung
9. Blood vessels
10. Heart
11. Elbow joint
12. Liver
13. Muscle
14. Kidney
15. Pancreas
16. Arm
17. Skin graft
18. Penis
19. Hip joint
20. Wrist joint
21. Finger joints
22. Leg
23. Knee joint

NERVOUS SYSTEM

THE nervous system is your communication and control system. It is rather like a telephone network that connects every part of your body to a powerful central computer – the brain. The telephone wires of your nervous system are the nerves themselves. They are made up of long, thin cells, called neurons, which are linked by connectors called axons to other nerve cells, and to muscle cells. They carry information from the skin, eyes, ears, and mouth, telling the brain what is going on outside the body. They relay instructions from the brain to the muscle cells concerning every movement that you make. They even control your heartbeat, blood sugar levels, and, by signalling to the brain when you need food and drink, feelings of hunger and thirst.

BRAIN

SPINAL CORD

NERVES RADIATE OUT FROM THE SPINAL CORD

ULNAR NERVE

MEDIAN NERVE

RADIAL NERVE

LUMBAR NERVE

SCIATIC NERVE

A touch too much
If nerves in the fingers detect pain, messages are sent up the arm to the spine, which is the main nerve cable to the brain. The messages are quickly processed in the brain and a signal is sent to the arm muscles to move the arm away — all in under a second!

BRAIN

SPINAL COLUMN

SPINAL CORD

VERTEBRAE

NERVES

The vital connection
Inside your spinal column lies a thick bunch of nerves, called the spinal cord, that make the connection between your brain and the rest of your body. Thirty sets of nerves emerge along the length of the spinal column to supply all parts of the body with nerves to make muscles work, or sensory nerves to bring messages back.

Protecting your nerves
The spinal nerves pass through a tunnel of bone in the middle of the vertebrae of the backbone; this protects them from injury. A broken spine often paralyzes all of the body below the break, because of the damage to the nerves. This is why it is important not to move an injured person after an accident.

34

Your communications network

THE brain and spinal cord are the centers of your communications network. From here, nerves radiate out, reaching every part of the body. Each nerve has thousands of individual nerve cells which carry electrical signals to and from the spinal cord and brain. These electrical signals are far too tiny to give you a shock, or harm you, but they are powerful enough to make a muscle move, or carry information from the sensory cells.

Most of the constant activity in your nervous system goes on without your being aware of it. But if you prick your finger your brain sends out urgent instructions, both to the motor nerves in your muscles to move the finger away, and to your mouth, lungs, and voice box to cry out in pain.

The nerves connect to allow information and signals from the brain to be relayed at the right time and to the right place. The brain can control complicated tasks by bringing into action many different nerve cells. Although it is only a little larger than a grapefruit, the brain contains millions upon millions of nerve connections, providing an unimaginable number of possible pathways and combinations, far more than any computer in existence. The brain can also do things that no computer is capable of, such as deciding between right and wrong, and making sense of the world around it.

Your on-board computer

THE nervous system can be compared to a computer system which is capable of interpreting, and responding to, things around it. The eye sees a glass of lemonade, or the nose smells a flower. Coded signals from the eye and nose flash to the brain where the codes are sorted out and matched to a series of stored memories: these tell the brain that lemonade tastes good and flowers look attractive in a vase. The brain computer then decides what to do – whether to drink the lemonade or pick the flower and the relevant muscles are given their instructions.

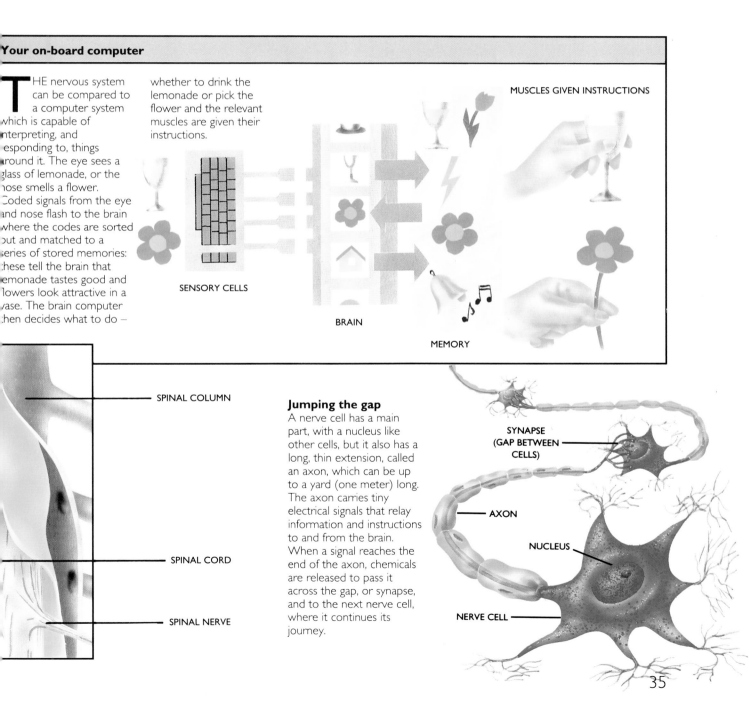

SENSORY CELLS

BRAIN

MEMORY

MUSCLES GIVEN INSTRUCTIONS

SPINAL COLUMN

SPINAL CORD

SPINAL NERVE

Jumping the gap
A nerve cell has a main part, with a nucleus like other cells, but it also has a long, thin extension, called an axon, which can be up to a yard (one meter) long. The axon carries tiny electrical signals that relay information and instructions to and from the brain. When a signal reaches the end of the axon, chemicals are released to pass it across the gap, or synapse, and to the next nerve cell, where it continues its journey.

SYNAPSE (GAP BETWEEN CELLS)

AXON

NUCLEUS

NERVE CELL

Your brain has a lot to cope with. Messages come in from your eyes, ears, and skin telling it what is happening outside your body. The brain makes sense of this information and decides what needs to be done in order to keep the body alive and healthy. Instructions go out from the brain to tell the muscles and organs what to do. The brain also listens to messages from inside the body controlling digestion, urine production, breathing, and heartbeat.

All this activity going on at once is rather like trying to ride a bike while singing, balancing a ball on your head, juggling, and wiggling your toes yet the brain does it easily and it manages to think as well. It does all this by setting aside different parts for different activities. These areas carry out their own jobs and send messages to each other when necessary, so that the whole brain works together.

Organizing your thoughts

YOUR brain is divided into different areas, each with a different job. As a result, the active part of the brain will change depending upon what you are doing. If you are looking at an object the visual area of the brain will be in use, but this may change if you think about the object instead. Then, the intellectual center at the front of the brain will become active.

In addition, the right and left sides of the brain have been shown to be different.

It is thought that feelings and imagination are part of the activities of the right side, whereas the left side may be responsible for the more logical areas of thought.

Scientists have discovered much of this by research done on people who have had strokes (a blood clot in the brain), or who have injured their heads in accidents. Often, when only a part of the brain has been damaged, only certain functions are affected. People with

Your sensitive body
Some areas of your body are more sensitive than others. There are more nerves in your hands, feet, and lips than in other parts of the body. As a result, more brain space is allotted to these areas. This drawing of a person is in proportion to the sensory nerves and shows how important these areas are.

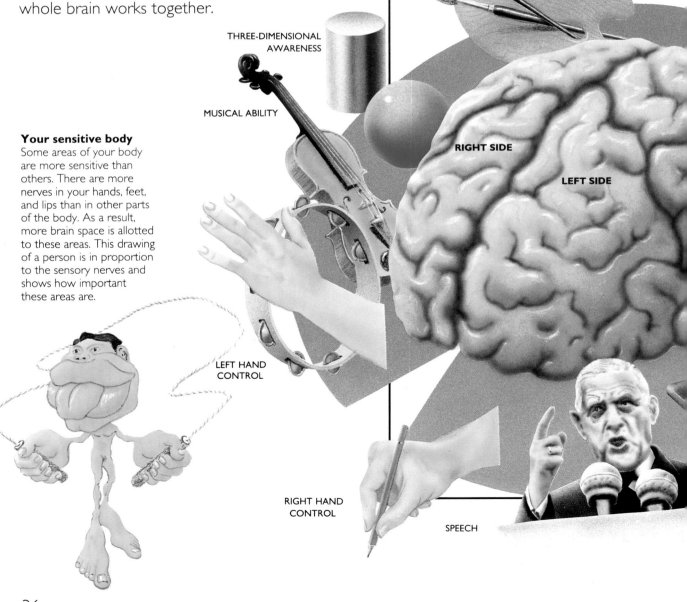

THREE-DIMENSIONAL AWARENESS

MUSICAL ABILITY

ARTISTIC ABILITY

RIGHT SIDE

LEFT SIDE

LEFT HAND CONTROL

RIGHT HAND CONTROL

SPEECH

damage to the right side may lose the ability to understand jokes, appreciate music, and sing. People with damage to the left side often cannot speak but can sing), cannot do mathematics, and lose the ability to recognize familiar faces. In some cases the damage allows a person to recognize their relatives by voice, but not by appearance. The brain is also split from front to back into areas. One part controls speaking, and large areas are set aside for controlling your muscles in activities like walking or writing.

SCIENTIFIC SKILLS

WRITING

CAL AND ATICAL SKILLS

The "map" in your brain

THE brain has specific areas to control the various muscles in the body. These parts of the brain are called the motor centers. Touch and the other senses are governed by areas of the brain known as the sensory areas. The motor centers and sensory areas are shown in the top diagram, side by side for simplicity.

You can see their actual locations in the bottom diagram.

The "mapping" of the motor centers and sensory areas in the brain has been worked out by observations during brain operations. Patients have undergone brain surgery while still awake – although they are given an injection to prevent their feeling any

pain! If the surgeon touches a sensory area in the brain then the patient feels a tingling sensation in the part of the body that matches the brain area. Similarly, touching one of the motor centers of the brain produces a movement in the relevant muscle in the body.

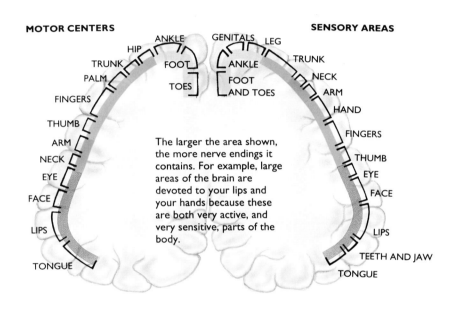

MOTOR CENTERS

ANKLE
HIP
TRUNK
PALM
FINGERS
THUMB
ARM
NECK
EYE
FACE
LIPS
TONGUE
FOOT
TOES

GENITALS LEG
ANKLE
FOOT AND TOES

SENSORY AREAS

TRUNK
NECK
ARM
HAND
FINGERS
THUMB
EYE
FACE
LIPS
TEETH AND JAW
TONGUE

The larger the area shown, the more nerve endings it contains. For example, large areas of the brain are devoted to your lips and your hands because these are both very active, and very sensitive, parts of the body.

Cutting the connection

The two sides of the brain are connected by a thick bundle of nerves in the corpus callosum. Sometimes the corpus callosum is damaged, or deliberately cut in order to cure brain problems, causing particularly unusual difficulties for the patient. For example, if he is asked to touch objects hidden behind a screen with the left hand, he cannot describe what the objects are – even though he can sketch what he thinks they look like with his right hand.

This is because signals from the left side of the body go to the right half of the brain. The speech centers lie in the left half

of the brain, and with no connection between the two halves, the patient is unable to say what he is feeling. However, signals from the left hand are reaching the side of the brain responsible for artistic and active ability, so the patient can draw the objects behind the screen.

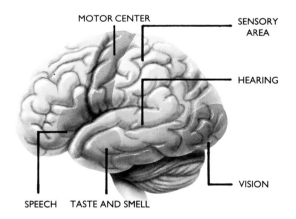

MOTOR CENTER

SENSORY AREA

HEARING

VISION

SPEECH **TASTE AND SMELL**

What you see is often not exactly what lies before your eyes. Your brain fills in the gaps for you. This computer picture is really a series of different colored squares, but your brain interprets it as a picture of a face.

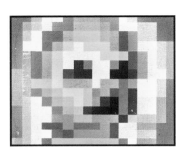

Choosing a chocolate

EVEN the simplest decisions require many different processes inside the brain. When you choose a chocolate the first thing that happens is that nerves in the eye send signals to the sight center in the brain, where the shape of the chocolate is analyzed. It is compared with other shapes seen before to decide whether it is a caramel or an orange cream. The brain then has to compare the memory of an orange cream with this new image, to decide whether or not the previous one was tasty, and whether or not to eat the new one. When the choice is made signals are sent to the motor centers, and the finger muscles reach for the chocolate. Eating it should stimulate the pleasure centers of the brain!

"A memory like a sieve!"

"I ALWAYS remember faces but not names." So how do you remember difficult things like telephone numbers, or a series of objects? The secret is to make them into a picture or story that is easier to recall. A top hat, an umbrella, a banana, a tennis racket and a monocycle can all be put into one image and remembered easily.

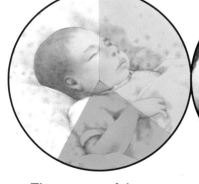

ASLEEP ☐   DREAMING ☐

The mystery of sleep

Why do you need sleep? No one knows for certain, but it seems that sleep is important for the brain, perhaps for recharging and sorting out memories. The patterns of sleep change throughout life. Babies need to sleep for more than half of the day, and nearly half of that time is spent in dream-type sleep. Adults spend only a third of the day asleep, and less time is spent dreaming.

Older people generally need less and less sleep, but their dream time remains about the same.

A trancelike sleep, or hypnosis, can be used to provide amusement, but hypnosis does have other uses, too. It can convince people to give up smoking or become more positive about difficult situations. This is because there is a part of the mind, called the subconscious, that controls your wishes and emotions. It is this area that is reached during hypnosis.

A sense of balance

YOUR ears don't just allow you to hear — they also help you to keep your balance. Without their balancing equipment you would fall over. Riding a monocycle, walking a tightrope, or simply standing up ll involve the brain sending onstant messages to the muscles in your legs to djust their posture.

Your sense of balance comes from three tiny half circles connected to a sac in the inner ear. The fluid in these semicircular canals moves around and stimulates nerves when the head tips. Since each half circle points in a different direction the brain can work out the direction of movement.

The movement of the fluid can be upset when you travel in a boat, or even if you spin around too fast, giving a feeling of giddiness and sometimes sickness.

Any sound — birdsong, someone playing a drum, or the noise of a jet plane — is simply a vibration which travels along the air waves. Your ears collect these sound vibrations and channel them into a tube full of sensory nerves, turning them into electrical signals that your brain can interpret. You can hear the difference between loud and soft sounds (volume), and high and low sounds (pitch). You can also tell from which direction a sound is coming, as your brain interprets the difference between the sound waves received by your two ears.

The loudness of a sound is measured in decibels. A sound that is just loud enough to hear is around ten decibels. If you are exposed to noise rated 80 decibels or higher, this could cause permanent damage to your hearing.

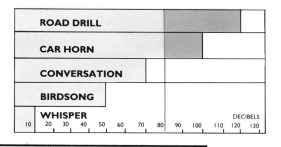

		DECIBELS
ROAD DRILL		
CAR HORN		
CONVERSATION		
BIRDSONG		
WHISPER		

10 20 30 40 50 60 70 80 90 100 110 120 130

How do you hear?

THE part of the ear that you can see (the pinna) is used to help find out from which direction a sound comes. The outer passage of the ear leads to the eardrum. This has a tight skin which picks up the sound and vibrates with it. The eardrum is attached to a bone called the hammer, which passes sound on to two more bones, called the anvil and stirrup because of their shape. The vibrations are then transmitted to the inner ear, where the sound is converted into electrical signals that your brain can interpret and "hear."

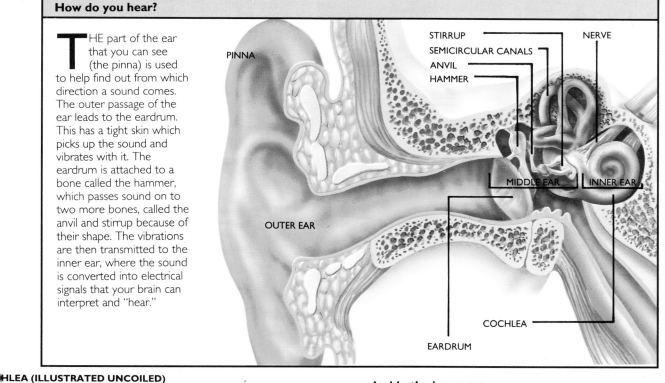

PINNA

STIRRUP
SEMICIRCULAR CANALS
ANVIL
HAMMER

NERVE

MIDDLE EAR   INNER EAR

OUTER EAR

COCHLEA

EARDRUM

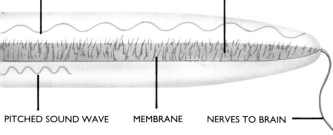

HLEA (ILLUSTRATED UNCOILED)

PITCHED SOUND WAVE   HAIR CELLS

PITCHED SOUND WAVE   MEMBRANE   NERVES TO BRAIN

Inside the inner ear

Once sound waves pass into the inner ear they are received in a coiled tube called the cochlea (it is shown uncoiled in the diagram for simplicity). The cochlea is filled with liquid which is vibrated by the action of the bones in the middle ear. The liquid vibrates hair cells attached to a membrane running along the center of the cochlea. The hair cells in turn pass the signal on to nerves which travel to the brain. A low pitched sound vibrates hairs farther up the cochlea than a high pitched sound, and so the brain can tell one pitch from another.

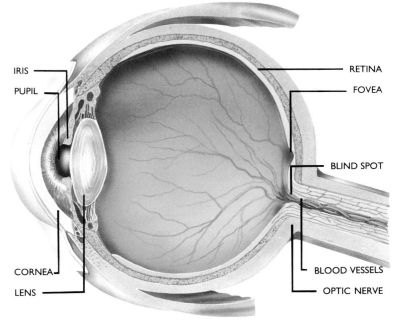

IRIS
PUPIL
CORNEA
LENS

RETINA
FOVEA
BLIND SPOT
BLOOD VESSELS
OPTIC NERVE

You use your eyes to make sense of the world around you. Imagine how difficult life can be for a blind person, and then you will get some idea of how important your eyes are to you. Simple tasks, such as crossing a street, walking up steps, or making a cup of coffee become difficult and dangerous without sight to guide you and locate objects.

How does the eye work?

LIGHT is gathered in by a curved transparent structure, called a lens, at the front of the eye. The lens bends the light so that a clear picture is formed at the back of the eye on the area called the retina. This is similar to the way in which a picture is created in a camera. Although you do not have removable film at the back of the eye, the retina does work rather like photographic film. It is made up of millions of light sensitive cells, similar to the light sensitive chemicals the film. Each cell is connected to a nerve which sends electrical signals to your brain. The nerves in your brain mak sense of the picture for you: edges are found, changes in color or shad are recognized, and the shapes of objects are picked out. So, although you may think that sight simple thing, many comp processes occur before can recognize an object admire a beautiful pictur

Speed of vision

Your vision changes as things move. The brain copes with this by making a series of pictures, hundreds of times a second. It puts them all together and you think that you see a continuous sequence. This is how television and films work too. There are hundreds of slightly different still photographs put together so quickly you see them as one sin movement. The first moving pictures of ten looked jerky because the process was not perfect. You can make your own moving pictures by draw a sequence of pictures n the edge of an old workbook and then flipp through the pages.

DID YOU KNOW?

Find your blind spot by looking at the left hand apple (covering your left eye) and moving the page forwards and backwards until the other apple disappears. This blind spot is the point on the retina where all the nerves in the eye connect to the brain, and there are no light-sensitive cells.

You see only a spot in the center of your vision in detail. There is an area on the retina with a high concentration of light sensitive cells (fovea) and this is the part of the retina you use when you look at something. Your brain "fills" in the rest from vague details.

At very high speeds, your eyes cannot detect movement because the light sensitive cells cannot recharge fast enough. Hunting animals like cats do not have such detailed vision as humans, but they can respond to rapid movements ten times faster.

Stand close to a mirror in a light place and close your eyes for a while. Open them and look quickly at the pupils in the center of your eyes. They shut up quickly to stop too much light getting into the eyes.

Seeing in color

YOU have three types of color-sensitive cells in you eyes: red, green, and blue. All the other colors are mixed from these combinations. Some peop have defects in their eyes so that one of the cell

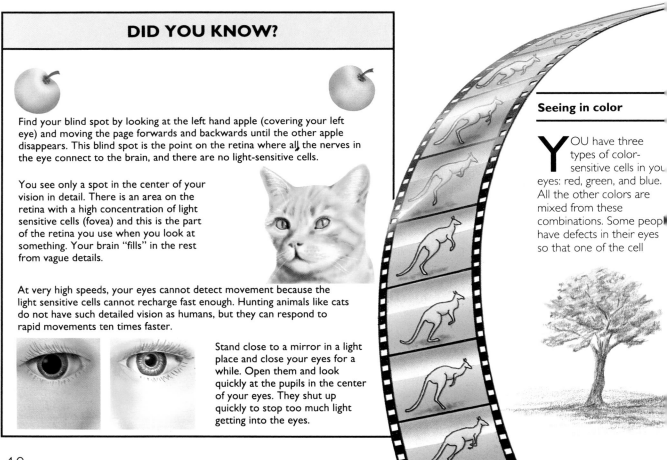

Optical illusions tell you something about the way in which your brain interprets things for you. When looking at this picture the brain tries to get the perspective right – unsuccessfully because the artist has drawn an impossible picture.

Distance and "stereoscopic" vision
How does your brain work out how far away an object is? One way is to compare the images of the object received by each eye.

These two pictures will be slightly different – very different if the object is close. This comparison can be used by your brain to compute distance. To test this, put one pencil on a

table and, looking along the edge of the table with only one eye, try to touch the tip with another pencil moving in from the side. Now try again with both eyes open.

...es is missing, called color ...dness". Tests for color ...dness use patterns to ...e familiar shapes. ...ple with red green ...r blindness cannot see ...red apples on the ...n tree.

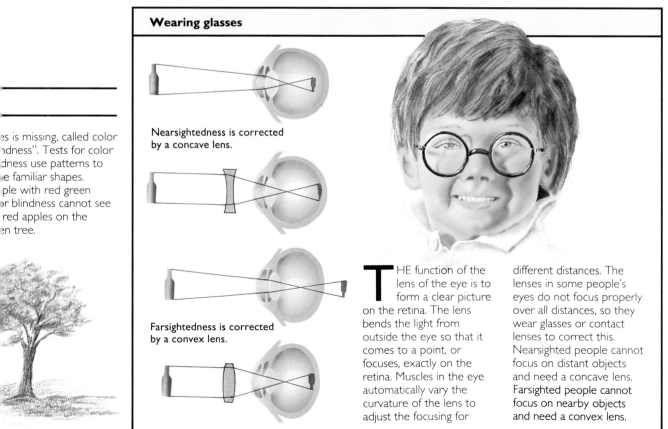

Wearing glasses

Nearsightedness is corrected by a concave lens.

Farsightedness is corrected by a convex lens.

THE function of the lens of the eye is to form a clear picture on the retina. The lens bends the light from outside the eye so that it comes to a point, or focuses, exactly on the retina. Muscles in the eye automatically vary the curvature of the lens to adjust the focusing for different distances. The lenses in some people's eyes do not focus properly over all distances, so they wear glasses or contact lenses to correct this. Nearsighted people cannot focus on distant objects and need a concave lens. Farsighted people cannot focus on nearby objects and need a convex lens.

The useful pituitary gland

The pituitary gland controls many different functions in the body. One of its jobs is to control the production and flow of milk to the breast when a mother is breast-feeding her baby. This is done by a hormone called prolactin.

PITUITARY

PROLACTIN

Fight or flight

What happens inside your body when you are frightened or angry? Nerves connected to the middle of the adrenal glands above the kidneys stimulate the release of adrenaline, a powerful, fast-acting hormone. Your body is immediately prepared for action; your heart rate goes up, breathing increases, blood is diverted to the muscles, and your skin becomes sweaty. These processes are often known as "fight or flight" reactions.

Brain
• controls the pituitary gland through the hypothalamus

Pituitary Gland
• the "master gland" controls many of the other glands and is connected to the brain at the hypothalamus

Thyroid Gland
• controls growth together with the pituitary gland
• also controls body heat and activity levels

Adrenal Glands
• control the balance of water and salt levels in the blood
• produce adrenaline, which prepares the body for sudden activity

Pancreas
• produces insulin and glucagon, which control sugar levels in the blood

Ovaries/Testes
• ovaries produce the female sex hormones estrogen and progesterone
• testes produce the male sex hormone testosterone

BRAIN
HYPOTHALAMUS

PITUITARY GLAND

THYROID GLAND

ADRENAL GLANDS

PANCREAS

FEMALE
OVARIES

MALE
TESTES

The body's chemical messengers

HORMONES, your chemical messengers, are manufactured in the endocrine glands. From these glands, the hormones are sent straight into the bloodstream and around the body. A master gland, the pituitary, controls many of the others. It lies beneath the brain and measures less than half an inch (one centimeter) across. In its turn, the pituitary is controlled by an area of the brain called the hypothalamus. In this way, all of the body's activities can be coordinated through the joint action of brain and hormones.

Over 20 hormones are produced by the endocrine glands, each controlling a different area or function of the body. The thyroid gland, together with the pituitary, makes hormones that control growth. The adrenal glands are also regulated by the pituitary, and make steroid hormones that control the build-up and growth of muscle and protein. Sugar levels in the blood are controlled by the hormones insulin and glucagon, both made in the pancreas.

Your built-in cl

INSIDE you is a built-in clock that regulates yo daily activities. A gland the brain, called the pine controls your clock. It secretes a substance call melatonin during the hou of darkness but stops during the day. Sleeping, eating, cell growth, and excretion are all controll on a day-to-day basis by

7 A.M.

amount of sex
ones present in your
changes throughout
d affects all areas,
cially the skin and hair.
t results of hormone
changes can include
ems with acne and
during puberty, and,
en, baldness as they
lder.
ne is caused by an
production of oil near
air root which clogs
e pores in the skin
fects the surrounding
Baldness occurs in
men as the level of
ale sex hormone,
sterone, increases.

SEBACEOUS GLAND

HAIR FOLLICLE

TRAPPED SEBUM

BACTERIA
INVADE

SKIN BECOMES INFLAMED

Dwarfs and giants

SOME people have
either very large or
very small amounts of
growth hormone and this
can cause them to be very
short or tall. The failure of
the pituitary gland to
produce enough growth
hormone leads to
"dwarfism." A tumor in the
pituitary can cause
overproduction of growth
hormone, and can lead to
the sufferer growing to
over eight feet (2.4 meters)
tall.

Your brain controls your
body by means of the nervous
system when it needs a quick
reaction. However, there are
many activities in the body,
such as the regulation of sugar
in the blood, or the amount of
food used up by cells
(metabolism), that do not
require quick action. It is these
activities that are controlled by
hormones. Hormones are
chemicals that flow around the
body in the bloodstream. Each
hormone has its own chemical
"address" which means that it
is recognized only by the cells
that need to respond to it. For
example, the hormones that control blood salts
have an "address," or chemical code, that is
recognized only by the kidney cells; and the
hormone insulin, which is made by the pancreas, is
recognized by the liver cells. It tells them to take
sugar from the blood and store it (see pages 34-
5). If the nerves are the body's telephone
communication system, the hormones are
its postal service.

internal clock.
u become aware of its
rtance if you travel a
distance into a
ent time zone. Your
clock is confused until
usts to the new times
eeping, waking, and
. This feeling is called
;

Nobody knows exactly
how body clocks work, but
it is known that both
animals and plants have
them. Flowers open, and
animals wake and become
active during their "day"
even if they are kept in

total darkness. Internal
clocks probably help
migrating birds to navigate
by the sun. Sea-dwelling
animals, such as oysters,

have clocks timed with the
tides. Their cycles are
maintained even if the
oysters are transported to
laboratories, well away
from the sea.

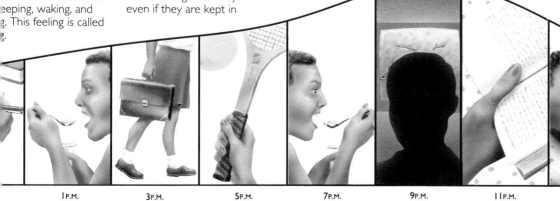

| 1 P.M. | 3 P.M. | 5 P.M. | 7 P.M. | 9 P.M. | 11 P.M. | 1 A.M. |

The skin is more than just an outside layer. It is the largest organ in your body, and it guards and protects you against all types of invaders and injury. It is also a sensitive organ, with thousands of nerve endings that send information back to the brain about what you are touching, how hot or cold you are, and whether you are in pain (see *pages 46-7*).

The tough outer layer of your skin is germproof and waterproof. Germs find their way into your body through the openings in your skin — your mouth and nose, and places where the skin is cut. Water is kept out — when you go swimming for example; and in — as body fluids which keep the organs inside you alive and working.

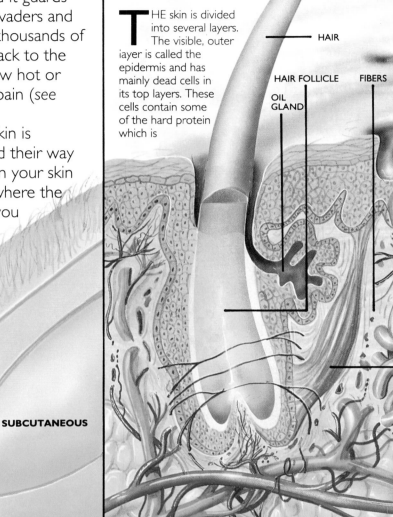

Your multilayered skin

THE skin is divided into several layers. The visible, outer iayer is called the epidermis and has mainly dead cells in its top layers. These cells contain some of the hard protein which is

HAIR

HAIR FOLLICLE

FIBERS

OIL GLAND

SUBCUTANEOUS

DERMIS

EPIDERMIS

WHORL

TENTED LOOP

ARCH

DOUBLE

ULNAR LOOP

RADIAL LOOP

ARCH

RADIAL

TENTED

WHORL

THE DIFFERENT COLOURS OF SKIN

The color of skin
The substance that colors your skin is called melanin, and is made by special skin cells where the epidermis meets the dermis. It protects the deeper layers of your skin against the harmful rays of the sun. Peoples from hot, sunny places need more protection than those from cooler places, so tend to have more melanin and be darker skinned. If lighter skinned people expose their skin to the sun then the increased production of melanin will make their skin "tan."

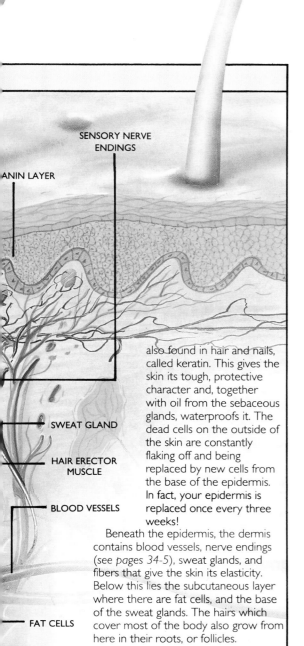

SENSORY NERVE ENDINGS

...ANIN LAYER

SWEAT GLAND

HAIR ERECTOR MUSCLE

BLOOD VESSELS

FAT CELLS

also found in hair and nails, called keratin. This gives the skin its tough, protective character and, together with oil from the sebaceous glands, waterproofs it. The dead cells on the outside of the skin are constantly flaking off and being replaced by new cells from the base of the epidermis. In fact, your epidermis is replaced once every three weeks!

Beneath the epidermis, the dermis contains blood vessels, nerve endings (see *pages 34-5*), sweat glands, and fibers that give the skin its elasticity. Below this lies the subcutaneous layer where there are fat cells, and the base of the sweat glands. The hairs which cover most of the body also grow from here in their roots, or follicles.

Healing a wound

If your skin becomes broken or cut, the flow of blood from the wound helps to clean it. The blood quickly clots, sealing the wound and preventing germs that normally live on the skin from penetrating inside. A scab forms across the wound, and the skin cells begin to grow back and fill the gap.

Long or deep injuries may need stitches to pull the sides together and encourage the skin cells to grow inwards to cover the wound.

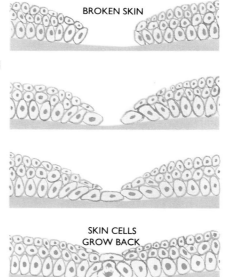

BROKEN SKIN

SKIN CELLS GROW BACK

The main temperature control in your body is provided by your skin. In hot conditions your skin flushes with blood, and the heat from the blood is conducted away from your body, making you cooler. You also sweat a salty liquid from the sweat glands in a layer of the skin called the dermis. The sweat takes heat away from your skin as it evaporates.

When you are cold the warm blood drains away from your skin to preserve its heat in the center of your body. You may find yourself shivering, because the rapid movements of the muscles create some extra heat.

Your unique fingerprints

The pattern of ridges on the underside of your fingertips is unique. Even identical twins have different fingerprints and, unless the skin is damaged below the epidermis layer, this pattern remains the same throughout life. This means that fingerprints are a useful means of identifying people, a fact that is often used by the police to catch criminals.

DID YOU KNOW?

You get "goose pimples" when you are cold because tiny muscles attached to the hairs in the skin pull the hairs upright. This might keep you warm if you were a bit more hairy, like your animal ancestors. As it is, it doesn't do much good!

Household dust is partly made up of flakes of dead skin.

Your skin wrinkles after being in the bath water too long because the hard top layer has softened and temporarily lost its waterproofing.

Only people and horses sweat properly. Dogs have sweat glands in their toes, but do most of their cooling down by panting.

The layer of skin called the dermis contains fibers which give the skin its elasticity and suppleness. As you get older this layer begins to break down, and the skin begins to sag and wrinkle, losing its youthful springiness.

All five main senses have been developed so that you can collect information about the world around you, particularly to help you survive. The information collectors are your eyes, ears, nose, mouth, and skin, but it is the brain that actually interprets what the information means (see *pages 39 and 40-1 for more about your ears and eyes*). Your senses of smell and taste are very closely linked, and were probably developed when humans had to hunt for food in an unfriendly environment and needed to distinguish between what was safe and what was poisonous. Your sense of touch will tell you to avoid dangers such as a hot plate, or a sharp knife, and it will also tell you if you are in pain.

A matter of taste

THERE are, surprisingly, only four main tastes: salt, sweet, sour, and bitter. It may seem difficult for all the various flavors of different foods to come out of these four. It is possible because the four tastes are usually blended, and because other factors sometimes affect your appreciation of food.

The texture of food often makes a difference, as well as its smell. In fact, many of the things that you may think you are tasting you are actually smelling. Try blindfolding someone and giving them an apple to eat while holding a ripe

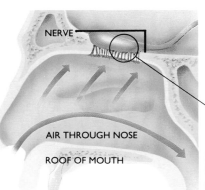

NERVE

AIR THROUGH NOSE

ROOF OF MOUTH

How do you smell?

It is true that humans do not smell very efficiently compared with most animals. Even so, your nose is capable of distinguishing around 4000 different substances.

Inside the nose are small areas of sensitive membranes which can detect chemicals dissolved in the wet mucus inside the nose. These smell "cells" send an electrical impulse to the brain, although exactly how this is done is not known.

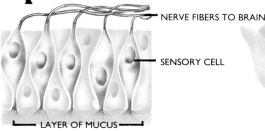

NERVE FIBERS TO BRAIN

SENSORY CELL

LAYER OF MUCUS

Cured by a pinprick

The ancient Chinese developed a method of medicine called acupuncture. This involves pushing very fine needles into the skin at certain points on the body, such as the hands or knees. Acupuncture has been used to relieve pain during operations, and might even be used instead of an anaesthetic to put people to sleep during operations. No one knows exactly how it works.

The importance of touch

YOUR skin is sensitive to many different kinds of situations. Heat, cold, pain, and pressure all provoke a different type of nerve cell in the skin. This provides the brain with useful information about the world outside the body.

Many of the messages sent to the brain by the nerve cells prevent you from damaging or injuring your body. Sensitivity to pressure and pain will stop you cutting yourself with a knife. Sensitivity to heat ensures that you don't burn yourself. Sometimes a sensation is felt as an itch, an insect or a piece of material irritates the skin, and then your brain commands your fingers to scratch.

HAIR

PAIN

TOUCH

COLD

HEAT

PRESSURE

...ear beneath their nose. They will probably be deceived by the smell, and think that they are eating a pear. This is why you cannot taste food when you have a cold – your nose is blocked and you cannot smell the aromas necessary for real enjoyment of food.

Nice and nasty smells
Why some smells are pleasant and some unpleasant is a mystery. It is certain that, for most people, the smell of a flower or the scent of a lemon makes them feel good; whereas the smell of compost or sewage makes them feel bad. These reactions may stem from humans' animal past when smells signalled good or bad foods.

Some businesses believe that good smells help people to work more efficiently, and pipe pleasant smells into their offices. But this may not really help because after a while the brain becomes accustomed to a smell and stops noticing it.

There are other senses apart from the five main ones. You know in what position your limbs lie without having to look because of "position" senses in your muscles. Your sense of hunger or thirst, controlled by hormones (see pages 42-3), will alert you to the fact that your body needs food or drink. And your sense of balance, controlled by the fluid in your inner ear, prevents you from continually falling over (see page 39).

The skunk protects itself by spraying its enemies with a foul-smelling liquid.

PLAN VIEW OF THE TONGUE

SWEET AND SALT

SOUR

SWEET

BITTER

SOUR

CROSS SECTION OF THE TONGUE THROUGH A VILLUS

ONE TASTE BUD

TASTE BUDS

SENSORY CELL

NERVES

NERVES LEAD TO SENSORY AREAS OF THE BRAIN

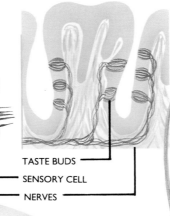

TONGUE

On the tip of your tongue
The surface of the tongue is covered by projections called villi, the sides of which contain taste buds. Before you can taste your food it must be dissolved by the fluid, called saliva, in your mouth. The taste buds are sensitive to chemicals in the food and signal to the brain whether the taste is sweet, sour, salty, or bitter.

Although all taste buds look identical there are four types, which are arranged in four areas on the tongue.

Each area detects one of the four tastes. Licking candy with the end part of the tongue is the best way to taste the sweetness, because all the sweet-sensitive cells are near the tip of the tongue.

47

REPRODUCTIVE SYSTEMS

ALL creatures must reproduce in order to avoid becoming extinct. This is why humans and animals have reproductive organs and why they experience both sexual and parental feelings. Human reproduction is an efficient process. Usually, only one or two children are born at a time, and they are carefully nurtured both before and after birth.

The information that your body needs to make a copy of itself is contained in your genetic program (see pages 52-3). This program designs your sexual organs so that they can bring a sperm and an egg together to make a baby. It also implants into your brain the desire to reproduce and be a parent. In this way, the human race continues.

Egg and sperm

Only one single sperm can fertilize the waiting egg. The successful sperm and egg normally unite after a man and woman have had sexual intercourse. Blood is released into the man's penis, enlarging it and

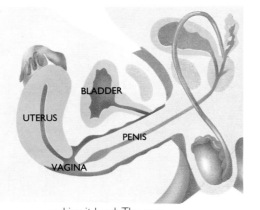

making it hard. The man slides his penis into the woman's vagina, and the movements of lovemaking produce orgasm. Male orgasm is the moment at which semen is squirted into the vagina.

The reproductive system: a means to an end

THE reproductive systems of men and women have one main function: the production of a fertilized egg. The male sexual organs must be capable of making sperm and delivering it to fertilize the egg inside the female, where the new baby will grow.

About 500 million sperm are made each day in the testes of a healthy, young man. They trickle down tubes to a collecting chamber called the epididymis, where they are stored for up to a week. Any sperm that is not used is absorbed back into the body.

The female system has a pair of ovaries, deep inside the body, which make eggs. Every month a single egg is released and passed down one of the fallopian tubes, ready for fertilization.

The sperm begin their journey to the

- BLADDER
- EPIDIDYMIS
- SPERM TUBE
- PENIS
- TESTIS
- ANUS
- RECTUM

female egg at the moment of ejaculation, when semen (a mixture of sperm and a sticky liquid) is squirted out of the man's penis by a pumping movement of muscles surrounding the base of the penis. Like tadpoles, sperm have little tails to help them swim. They move into the woman's vagina, through the entrance to the uterus (womb) and into the fallopian tubes where an egg may be waiting to be fertilized.

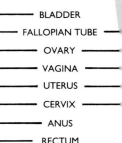

- BLADDER
- FALLOPIAN TUBE
- OVARY
- VAGINA
- UTERUS
- CERVIX
- ANUS
- RECTUM

Contraception - preventing fertilization

CONTRACEPTIVES allow people to make love without having to become pregnant. Barrier methods such as the condom and the diaphragm prevent sperm swimming beyond the vagina. The condom fits over the penis, and the diaphragm fits over the cervix. Intra-uterine devices (IUD's) are fitted into the uterus and prevent an egg implanting in the wall. The pill contains hormones that mimic the body's own hormone levels to interfere with either egg production, or the implanting of the embryo.

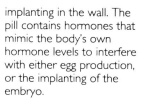

CONDOM

DIAPHRAGM

IUD

CONTRACEPTIVE PILL

From egg to embryo

NOT all the sperm will survive the journey through the uterus and up the fallopian tube to meet the egg. Those that do swim around the egg, until just one of them breaks through its thin outer wall. The egg then changes its outer wall so that no more sperm can enter.

The sperm and the egg join into one, and growth begins. The single cell divides into two, then four, and so on, until after about five or six days a ball of a hundred cells is formed, and attaches itself to the wall of the uterus.

0 hours
Fertilization: one sperm breaks through the thin, outer wall of the egg

30 hours 2 cells: the single cell divides into two

4 days Ball of cells: the ball of cells has moved down the fallopian tube to the uterus

5 days
Implantation: the ball of cells is now implanted into the uterus wall

The monthly cycle

Unlike men, who produce sperm continuously, fertile women produce an egg only once every month. The release of the egg is controlled by hormones called progesterone and estrogen (see page 42) that cause changes in the woman's body, especially in the ovaries and the uterus.

The cycle starts when an egg in one of the ovaries is mature and ready for release. It moves out of its ovary into a fallopian tube. In preparation for possible fertilization the lining of the uterus becomes thick, ready to receive a fertilized egg.

If the egg is not fertilized it dies, and the lining of the uterus is released as a monthly "period." The cycle can then start all over again.

BLEEDING   EGG DEVELOPS IN OVARY   EGG RELEASED INTO FALLOPIAN TUBE   EGG MOVES TOWARDS UTERUS

UTERUS WALL THICKENS, READY TO RECEIVE THE EGG

WEEKS   1   2   3   4

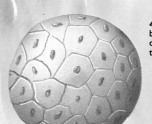

49

From embryo to birth

4 weeks
The embryo is mostly made up of brain and backbone.

6 weeks
The fetus begins to develop arms and legs.

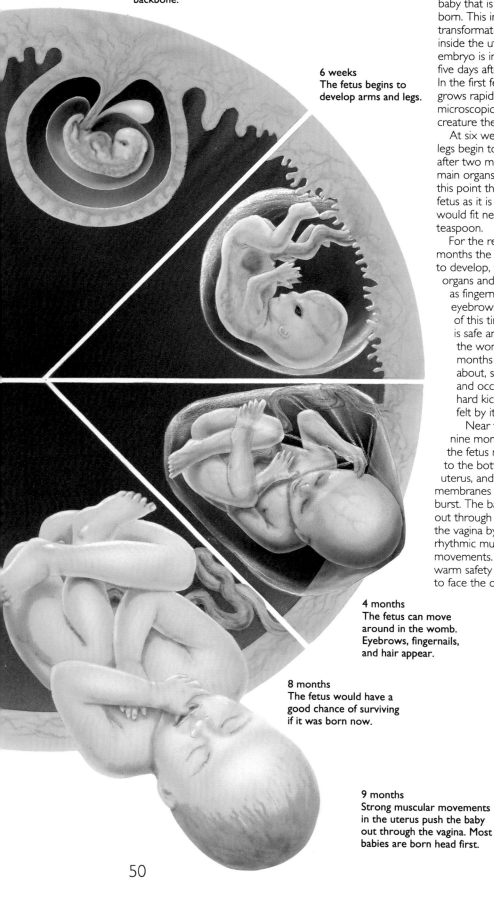

I

T TAKES nine months for an embryo to develop into a full-grown baby that is ready to be born. This incredible transformation takes place inside the uterus where the embryo is implanted about five days after fertilization. In the first few weeks it grows rapidly, from a microscopic dot to a creature the size of a pea.

At six weeks arms and legs begin to appear, and after two months all the main organs are formed. At this point the embryo, or fetus as it is now called, would fit neatly into a teaspoon.

For the remaining seven months the fetus continues to develop, with sexual organs and details such as fingernails and eyebrows appearing. All of this time, the fetus is safe and warm inside the womb. After three months it can move about, suck its thumb, and occasionally give a hard kick which can be felt by its mother!

Near the end of the nine months pregnancy the fetus moves down to the bottom of the uterus, and eventually the membranes surrounding it burst. The baby is pushed out through the cervix into the vagina by strong, rhythmic muscular movements. It leaves the warm safety of the womb to face the outside world.

4 months
The fetus can move around in the womb. Eyebrows, fingernails, and hair appear.

8 months
The fetus would have a good chance of surviving if it was born now.

9 months
Strong muscular movements in the uterus push the baby out through the vagina. Most babies are born head first.

Two way twins

T

HERE are two different ways of producing twins. Identical twins are formed from one fertilized egg that splits into two. Two babies are produced, both with the same genes, so they will look almost identical as they grow up. Nonidentical twins are produced when two eggs are released from the ovaries in the same month, and both are fertilized. The embryos are formed from different sperm and eggs so they will look different, and can be different sexes. Identical twins will always be of the same sex, and will always have the same colored eyes. Nonidentical twins can have different colored eyes.

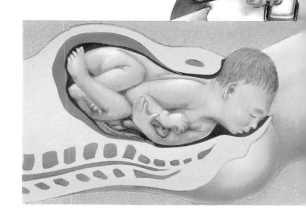

50

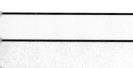

NIDENTICAL TWINS

idential twins can have rent colored eyes.

NTICAL TWINS

Identical twins will always have the same colored eyes.

While it is inside the uterus, the baby is given oxygen and nourishment through the umbilical cord. This links the fetus to the placenta, which surrounds the baby and connects it to the mother's blood supply.

PLACENTA ——————

UMBILICAL ——————
CORD

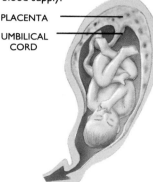

Ultrasound: checking the baby

Modern technology makes it possible to look at the fetus while it is still inside the mother. A machine that sends out very high frequency sound waves is used to obtain pictures on a television screen. The system works rather like radar; sound waves are bounced off the fetus, and the machine receives the echoes which it converts into a picture. The fetus can be checked for abnormalities, and the parents can find out whether it is a boy or a girl.

AN ULTRASOUND PICTURE

It is amazing that a tiny embryo, smaller than the period at the end of this sentence, can carry within it all the necessary information to make a new person. This miracle of self-assembly is possible because the fertilized egg contains an "instruction manual," in the form of genes, which tells it how to make a new human being. Half of your genes are inherited from your mother, half from your father, so you will have some of the characteristics of both parents.

As the embryo grows, instructions are passed on by the genes to each new cell, so that every cell knows how to fit in with the rest. The genes also control the formation of cells, in order to make all the different tissues and organs needed for a complete human being.

Breast-feeding

Breast-feeding is the natural way to feed the baby. A mother's milk contains an ideal mix of nutrients, and is also thought to contain antibodies that can protect the baby against illness. The mother's breasts start to increase in size during pregnancy.

The baby's sucking movements stimulate the milk production, and the milk can continue until the baby starts taking solid food, or bottled milk.

DID YOU KNOW?

Before birth, a baby's heart has a bypass that prevents most blood from going to the lungs. At birth the bypass is shut, and blood rushes to the lungs, ready to take oxygen in.

The testes hang outside the body to keep them cooler than the other organs. This is because sperm production is reduced if the testes get too hot.

It is possible to remove eggs from a woman and fertilize them with sperm in a laboratory. This technique can be useful if the woman has blocked fallopian tubes, or other problems which make her infertile. The embryo can then be put into the mother as if fertilization had happened normally. Babies born by this method are often called test-tube babies, even though test tubes are unlikely to be used in the process.

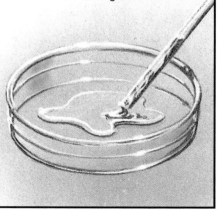

Hidden inside you there is an "instruction manual," rather like the program of a computer, that regulates and directs your body. This information is stored in the nucleus of every cell, telling it what it should do and giving it different instructions depending on what sort of cell it is. This book of life is not written on paper but on a long threadlike substance called DNA. The manual is divided up into short chapters called genes. Each of the genes gives instructions to a particular part of your body. There are, for example, genes that say what color your eyes will be. Other genes have the instructions for the red cells that carry oxygen around in your blood. When all your genes are put together they make a complete instruction manual for one person.

A clone is a genetically identical group of cells or creatures. Most of the cells in your body are clones of each other, because they have the same genes, even if they do not use the genes in the same way. Identical twins are clones of each other. The idea that a whole new person can be cloned from a few cells is unlikely ever to come true.

Damaged DNA

If the DNA in a cell is damaged, the genetic program cannot work properly. This causes a form of illness called genetic disease. There are many sorts of genetic problems, ranging from the fairly common Down's syndrome to the less well known neurofibromatosis, a disease which causes growths on the skin. The boy in the wheelchair (right) suffers from muscular dystrophy, a wasting disease of the muscles which mainly affects boys. Hemophilia, sickle cell disease, cystic fibrosis, many types of mental retardation, and some aging diseases have all been shown to be controlled genetically. As more is discovered about genes, more problems are found to be genetic in origin.

Your instruction manual

YOUR genetic program is written in a substance called DNA which is arranged in extremely long, thin spirals. The DNA contains letters which make a language. The words and letters are different from our language, but the principle is the same. The cell can read it even if you cannot.

The program written in the DNA is the key to life. It determines whether you have fair hair, long legs, a tendency to heart disease, a small nose, and what sex you are – as well as many other characteristics. It does this by programing you through your cells in the same way that a floppy disk programs a computer.

The message written in the DNA programs all living creatures, from bacteria to birds. In fact, most creatures look very similar when they begin life. It is the DNA, the genetic program that makes the difference between a sea urchin and a human.

This requires a huge amount of information which is all stored in the DNA. Computer disks may be good at information storage, but DNA is far

DNA PROGRAM

COMPUTER PROGRAM

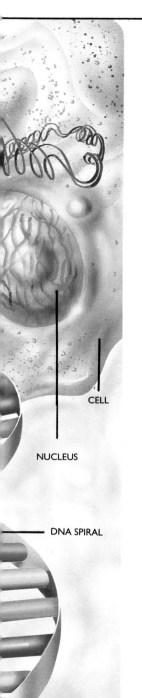

CELL

NUCLEUS

DNA SPIRAL

better. Inside a tiny cell nucleus, DNA can store one hundred times more information than a desk top computer. DNA can store this amount of information because it is so long and thin. Each cell contains about two yards (two meters) of DNA, which means that we each have about 16,800 million miles (27,000 million kilometers) of it: enough to go around the Earth 700,000 times!

Genes are passed on from parents to children, but only half of the genes from each parent go to the new child, and not always the same half. This means that everyone is different genetically, and that although sisters and brothers share some genes they never share exactly the same ones (unless they are identical twins).

There are probably around a hundred thousand genes inside you. The exact number is not known because researchers have only managed to find out about a few thousand. So there is a long way to go before anyone will understand your genetic program and with it such problems as ageing, cancer, or how an embryo develops.

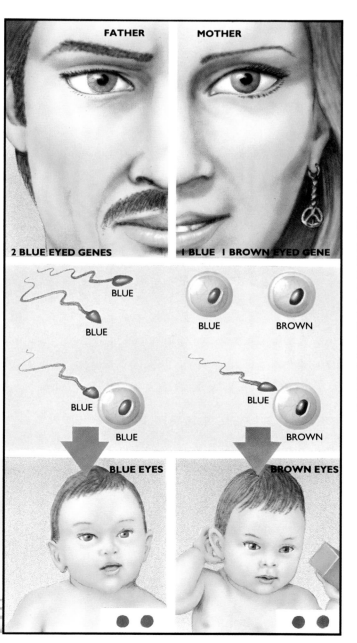

FATHER   MOTHER

2 BLUE EYED GENES

BLUE

BLUE

BLUE

BLUE

BLUE

I BLUE I BROWN EYED GENE

BLUE   BROWN

BLUE

BLUE   BROWN

BLUE EYES   BROWN EYES

Whose eyes?
Eye color can be used to illustrate the way that genes are passed on from parent to child. Everyone has two genes controlling eye color, either brown or blue. If you have one of each then the brown eye gene will be dominant, and you will have brown eyes.

If a brown eyed woman with one brown and one blue eye gene marries a man with two blue eye genes they could have children with either blue or brown eyes. But in this case it would always be the mother who determined the color of the children's eyes, depending upon which gene, the blue or the brown, she passed on. Most characteristics — height, weight, hair color, skin type — are controlled by many different genes, so their inheritance is more difficult to trace.

DISEASES AND MEDICINES

THERE are two main types of disease that can affect your body — those which are caused by an infection from outside, and those that happen because of a breakdown inside. Diseases such as cancer, rheumatism, and arthritis as well as mental disorders all belong to the second type. Common childhood diseases such as mumps, chicken pox, and measles are the result of an attack from outside the body, and belong to the first type.

Your body is constantly being attacked by germs (bacteria or viruses), or tiny animals which can live and grow inside the body, called parasites. You survive because your immune system (see pages 26-7) detects the invaders and sends antibodies to attack them.

Useful antibiotics
Antibiotics are used to fight off harmful bacteria in the body. They can cure many diseases, such as pneumonia and tuberculosis, and infected wounds. However, sometimes the bacteria become resistant to the antibiotic, and it will not work. Sometimes the antibiotics can kill off the useful bacteria that live in your body.

What are you frightened of?
Many people suffer from dizziness when they look over a cliff, or down from a tall building. This feeling is called vertigo. Sometimes vertigo can make people frightened when there is no actual danger. This sort of fear is called a phobia and can take many forms.

Why vaccinate?
Vaccination gives you protection against many diseases caused by germs (see pages 26–7). Vaccination is especially important for fighting viral diseases, because your immune system is your only defense against viruses.

Fear of heights is called acrophobia, whereas agoraphobia is a fear of open spaces. Other common phobias are claustrophobia, which is a fear of being shut in, herpetophobia, a fear of snakes, and aerophobia, a fear of flying.

AIDS
AIDS is a disease caused by a virus called HIV. There is not, as yet, any cure. The virus attacks the body's immune system so that the victim becomes unable to fight off infection and diseases.

The HIV virus is usually passed on through infected blood, often during sexual intercourse, or by drug addicts who share dirty needles to inject themselves. Sensible precautions, such as wearing a condom during intercourse, can slow the spread of AIDS.

Malaria: a disease of the tropics
Malaria is caused by a single celled animal, called a plasmodium, which attacks human blood and liver cells, eating them and producing poisons that cause a fever. The disease is passed on by mosquitoes as they go from person to person, biting and sucking blood. Hundreds of millions of people in tropical countries suffer from malaria. Many of these people will die of the disease, while others suffer for years as the disease comes and goes.

Some viral diseases

Viruses are smaller than bacteria and attack the cells from inside. Antibiotics have no effect against a viral infection, so your immune system must fight the invading virus alone. Colds, influenza, chicken pox, mumps, measles, cold sores, glandular fever, meningitis, and shingles are all caused by different viruses.

Painful Joints

Arthritis is a painful disease of the joints. In one type, the bone and cartilage become damaged over the years due to wear. In another, arthritis develops when the body's own defense system causes the membrane in parts of the joints to break down and the joints become inflamed. In both cases, moving the affected joints is difficult and painful.

Sometimes the attacking germ or parasite is so harmful that your immune system is not strong enough to fight it on its own. This is where modern medicine can help. Vaccinations prevent people from catching dangerous diseases, and drugs such as antibiotics attack the invading germs with powerful chemicals. Thanks to a world-wide vaccination campaign a disease called smallpox, which used to kill millions of people, has now disappeared. Other more common diseases, such as tuberculosis, diphtheria, and whooping cough, can now also be treated.

Curing cancer

Cancer is caused when one of the cells in the body becomes abnormal and divides uncontrollably, gradually taking over an area of the body. It is important to detect and treat all cancers at an early stage. Modern surgery includes the use of lasers to burn away the cancerous cells. In the future, research will concentrate on finding chemicals that can destroy the cancerous cells without harming healthy ones.

Your transparent body

If you damage a joint or bone you can find out what has happened by having an X-ray. This is a photograph of the inside of the body. The X-rays penetrate through the soft tissues of the body, but cannot go through the hard tissues of the bones. These show up as light patches on the dark X-ray film. You could say that humans are transparent to X-rays, but not to ordinary light.

Baron von Münchausen

Baron von Münchausen was an eccentric German soldier who told many fantastic tales. A type of mental problem has been named after him. People with Münchausen's

syndrome fake illness in order to be admitted to the hospital. They may use imitation blood capsules, take insulin, or injure themselves on purpose. One patient forced an airplane to land after complaining of chest pains. Others have persuaded surgeons to amputate (cut off) limbs after only minor injuries!

All humans have a genetic program that controls how they develop and grow (see *pages 52-3*). In a healthy environment, with a varied diet, and love and care from parents and friends, most people have the potential to achieve a happy and useful life.

Everyone passes important milestones in their development. One stage builds on another, so that a baby will learn to crawl before it can walk, and to walk before it can run. The human body grows rapidly during childhood and up to puberty, but gradually slows down until growth stops around 20. No one understands exactly how your body knows when to stop growing, but it may be that "time clocks" in your genes tell your cells when to stop dividing.

The story of growing up

ONE of the first things a baby can do is cry. Later it starts to look around and recognize voices and faces, and within a few weeks can smile at its parents. Learning to crawl and walk usually happens between one and two years. Other milestones, such as talking, reading, and writing, are all perfected throughout childhood.

Puberty is a time when many changes occur in the body. For girls, puberty

How long will you live?
In most Western countries the average life expectancy is around 76 years for women and 73 years for men. But there are some ways in which you can control your chances of living to an old age. You will receive plus points for not smoking, driving carefully, eating a varied, nutritious diet with little fat, and taking regular exercise. Minus points are awarded for drinking too much alcohol, although some research suggests that people who drink moderately seem to have a slightly longer lifespan than those who do not drink at all.

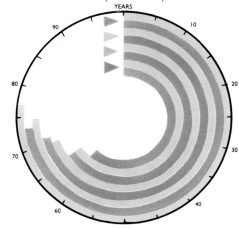

▶ AVERAGE
▷ OVERWEIGHT
▷ HEAVY DRINKERS
▶ SMOKERS

◼ MEN ◻ WOMEN

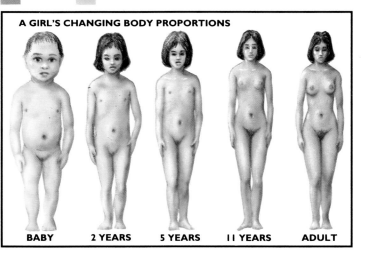

18 YEARS

2 YEARS

5 MONTHS

25 YEARS

40 YEARS

A GIRL'S CHANGING BODY PROPORTIONS

| BABY | 2 YEARS | 5 YEARS | 11 YEARS | ADULT |

Your changing shape
The shape of your body changes as you grow older. Babies have much bigger heads, compared with their body length, than adults do. At birth, the head takes up one quarter of the total body length, and by six years the brain is almost a full adult size. Your body grows up in spurts, mostly in the first year, between five and seven years and at puberty (11-16 years). After puberty you are more likely to get fatter than grow taller. Unless you overeat, your body will keep roughly the same proportions until old age, when it begins to shrink again.

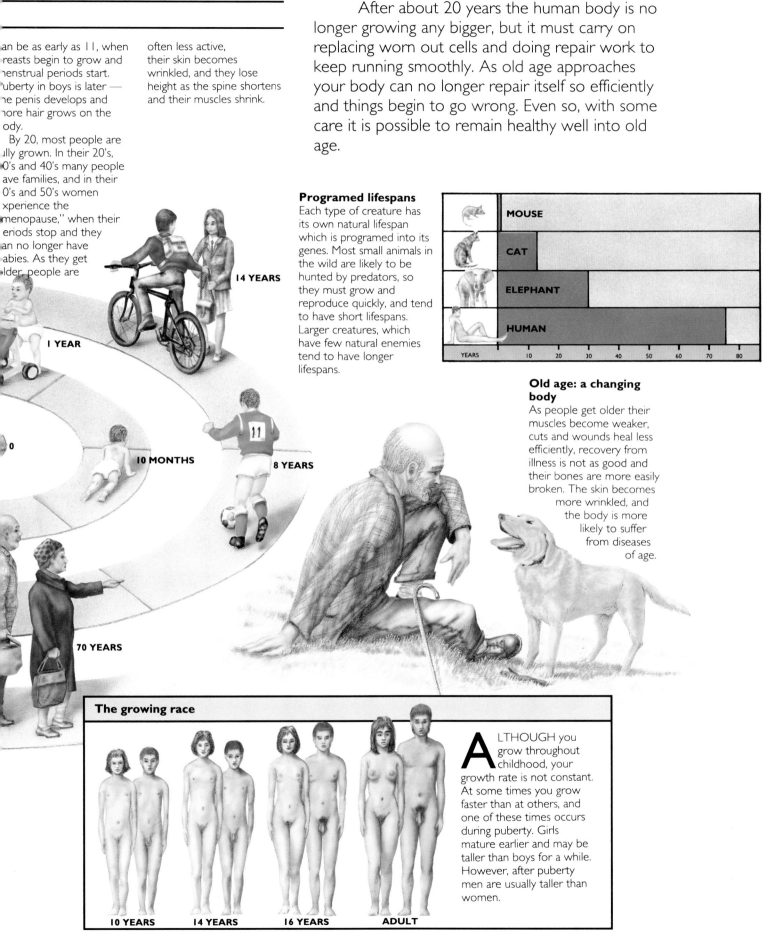

an be as early as 11, when
reasts begin to grow and
henstrual periods start.
'uberty in boys is later —
he penis develops and
hore hair grows on the
ody.
By 20, most people are
ully grown. In their 20's,
0's and 40's many people
ave families, and in their
0's and 50's women
xperience the
"menopause," when their
eriods stop and they
an no longer have
abies. As they get
•lder, people are

often less active,
their skin becomes
wrinkled, and they lose
height as the spine shortens
and their muscles shrink.

After about 20 years the human body is no
longer growing any bigger, but it must carry on
replacing worn out cells and doing repair work to
keep running smoothly. As old age approaches
your body can no longer repair itself so efficiently
and things begin to go wrong. Even so, with some
care it is possible to remain healthy well into old
age.

Programed lifespans
Each type of creature has
its own natural lifespan
which is programed into its
genes. Most small animals in
the wild are likely to be
hunted by predators, so
they must grow and
reproduce quickly, and tend
to have short lifespans.
Larger creatures, which
have few natural enemies
tend to have longer
lifespans.

MOUSE		
CAT		
ELEPHANT		
HUMAN		

YEARS 10 20 30 40 50 60 70 80

**Old age: a changing
body**
As people get older their
muscles become weaker,
cuts and wounds heal less
efficiently, recovery from
illness is not as good and
their bones are more easily
broken. The skin becomes
more wrinkled, and
the body is more
likely to suffer
from diseases
of age.

14 YEARS

1 YEAR

0

10 MONTHS

8 YEARS

70 YEARS

The growing race

10 YEARS   14 YEARS   16 YEARS   ADULT

ALTHOUGH you
grow throughout
childhood, your
growth rate is not constant.
At some times you grow
faster than at others, and
one of these times occurs
during puberty. Girls
mature earlier and may be
taller than boys for a while.
However, after puberty
men are usually taller than
women.

BODY FACTS

Why do boys' voices deepen at puberty?

The male hormone, testosterone, controls many of the changes in a boy's body at puberty. One of its effects is to make the larynx much larger, and the vocal cords almost twice as long, deepening the voice.

What happens when food "goes down the wrong way?"

The food has touched a sensitive part of the throat which makes you cough to stop the food going into the lungs.

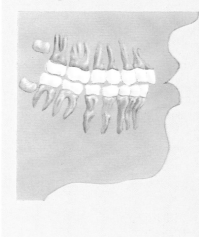

What are wisdom teeth?

Your wisdom teeth are teeth at the back of the mouth that don't normally appear until the late teens or early twenties. They are called "wisdom" teeth because they appear so long after all the other teeth, when people are supposedly older and wiser.

Why do people snore?

People who snore often breathe through their mouths, because the nose or throat is blocked. As air is taken in through the mouth it vibrates a flap of skin at the back of the mouth called the uvula, causing the loud, rattling noise of a snore.

Why do you yawn?

After long periods of shallow breathing — which happen if you are tired, or if you sit still for a long time — your body often needs more oxygen. Yawning is a reflex action to force more air into the lungs.

Do cells live for ever?

Red blood cells live for only four months and are continually being replaced, whereas nerve cells must last for life because they cannot be replaced.

Are germs always harmful?

Your digestive system is full of microbes living happily off the food you eat, causing no problems and even making useful vitamins in some cases. In fact, antibiotics taken to kill harmful bacteria can sometimes kill the useful ones as well, causing more problems than they solve.

What are freckles?

Freckles occur when the coloring in the skin — called melanin — is not even, but appears as tiny spots. More freckles will appear as the skin is exposed to the sun to protect the skin from the sun's harmful rays.

Is your hair alive?

From the moment your hair grows out of its root, or follicle, it is dead — its cells cannot divide to produce new hair. You lose, on average, 75 hairs every day from your head but these are continually being replaced by new hair.

What causes pins and needles?

If the blood supply is slowed down in one part of the body the nerves stop working and may be numbed. When the blood supply returns you will feel the pricking sensation of "pins and needles" as the nerves begin to work again.

Is your heart heart-shaped?

No. Your heart is the size of a clenched fist and shaped rather like a pear, with its broad upper part to the right, and the thin end to the left.

ABOUT THE BODY

Of your total body weight, the skin takes up 16 percent, your muscles about 40 percent, your bones around 25 percent, and your brain 2 percent.

Your body is between 55 percent and 70 percent water, with approximately another 10 percent chemically combined in fats, proteins, and carbohydrates. Men have a higher proportion of water in their bodies than women do, and children have a higher proportion than men.

An adult has about 17½ pounds (8 kilograms) of fatty substances in his or her body.

Of your 5 quarts (6 liters) of blood, over half is in the veins, and a fifth in the lungs.

The most common element in the body, found in many different forms, is oxygen (65 percent – enough to fill 200,000 cans of cola). There is enough carbon to light 30 barbecues, enough phosphorus for about 20 boxes of matches, enough hydrogen to fill a 9000 quart (10,000 liter) balloon, and the same amount of iron as a small nail. The remaining salts would fill a bag of fertilizer big enough to bring a large lawn to life!

You have 300 bones at birth, but only 208 as an adult. This is because some of the bones join together as you grow up.

You have 639 muscles. The longest muscle is the sartorius in the leg, and the shortest is in the inner ear, the stapedius.

RECORD-BREAKING HUMANS

The tallest man on record was 8 feet (2.47 meters) and the shortest was 2 feet (67 centimeters).

•

The most common infectious disease is the cold and the most common noninfectious one is gingivitis, which causes gum swelling and may affect up to 80 percent of people.

•

The record time anyone has been known to live without their heart beating is held by a Norwegian. He fell into icy water, and survived there for four hours before being pulled out and revived on a heart-lung machine.

•

The longest attack of hiccups was 67 years, and the longest bout of sneezing lasted for over two and a half years, with about a million sneezes in the first year.

•

The loudest snore ever recorded was 90 decibels. It broke the local bylaws in Vancouver, where the maximum permitted noise level was 80 decibels.

•

The longest-living heart transplant patient has survived 18 years since the operation. The first heart/lung/liver transplant was performed in 1987, and a five organ transplant was performed on a three year old who had never eaten solid food. The liver, lungs, pancreas, small intestine, large intestine, and part of the stomach were transplanted.

•

The longest swim with stops was 1,825 miles (2,938 kilometers) down the Mississippi River, in 742 hours. The longest unbroken swim was 299 miles (481 kilometers) in 84 hours, 37 minutes.

GLOSSARY

Antibiotics: Chemical substances, taken as medicine, that are capable of destroying harmful bacteria.

Antibodies: Chemicals that find and kill germs. They are made by the B-cell type of white blood cell.

Arteries: Tubes which carry blood away from the heart. Most arteries transport the oxygen-filled blood to the rest body, but the pulmonary artery, which runs from the heart to the lungs, carries deoxygenated blood in it.

ATP (Adenosine triphosphate): The name given to the energy-storing chemical made in each cell by the mitoc ATP packets can be used by all parts of the cell to provide energy for all its processes.

Bacteria: Microbes that live inside the body. Some cause disease (for example: Legionnaires' disease, tetanus, gum or gingivitis, cholera, and typhoid), but most are harmless and help the body to recycle waste and even, in some case vitamins. Bacteria are one celled, and are just visible with an ordinary microscope.

Bile: A greenish-yellow liquid made in the liver. It is stored in the gall bladder and is released into the intestine where helps to dissolve fats.

Blood: A reddish-purple liquid made up of a watery fluid (plasma) with cells (white and red) floating around in it. Th blood is the body's transport system to the cells.

Blood vessel: Any of the tubes containing and carrying blood around the body (arteries, veins, and capillaries).

Capillaries: Tiny tubes which carry blood through the organs of the body. Capillaries have thin walls so that the sug oxygen carried by the blood can be easily off-loaded to the cells.

Carbohydrates: Foods, such as starch and sugars, which provide energy.

Carbon dioxide: A gas which is produced as sugars are turned into energy. It is poisonous to the body if too much up, and is carried away in the blood.

Cell membrane: The thin outer wall of a cell which controls what goes in and out of the cell.

Cells: The tiny units which make up all the organs and tissues of the body. They are organized to work on their own still rely on a food supply, and waste removal, from the blood. Cells usually have a specialized job to do in the body.

Chromosomes: Long threads made of DNA that are found in the nuclei of cells, and have the genes strung out alo them. We have 46 chromosomes in each of our cells, except for sperm and eggs which have 23.

Decibel: A measure of the loudness of sound. The sound of a pin dropping is about 10 decibels. The sound of a pn drill is about 120 decibels..

Digestion: The breaking down of food into particles that can dissolve in the blood.

DNA: DNA is the substance that makes up your genetic material — the instruction manual that is used by your cells build your body. DNA is shaped like a very long, thin ladder twisted up into a spiral. The rungs of the ladder can be c types, and these rungs are used as a language to write the genetic instruction manual.

Egg: A female cell, produced once a month, which can form a new person if it is fertilized by a male sperm. When th happens half the chromosomes from the mother (in the egg) and half from the father (in the sperm) make up the ful genetic instructions from which the embryo develops.

Embryo: The name given to the developing baby during the very first stages, from the first division of the fertilized e the main organs are formed and working.

Endocrine gland: A gland which manufactures hormones for release into the bloodstream.

Enzymes: Complicated chemicals which perform all the important jobs in the body, such as digesting food, turning f into energy, and making new structures and cells. Enzymes are a special variety of protein.

Excretion: The removal of waste substances, such as urine, from the body.

Fats: Foods that are highest in energy. Butter, oils, lard, and meat fat are almost pure fat.

Fertilization: The joining together of male sperm and female egg to make an embryo.

Fetus: The stage in development of a baby after about eight weeks, when the limbs and main organs are formed, un birth. Before this time it is called an embryo.

Follicles: The roots in the skin from which hairs grow.

Genes: The instructions passed on from parents to their offspring which determine their looks and characteristics. Th made of DNA, and are carried in the chromosomes which are contained in the nucleus of each cell.

Germs: Microbes that are harmful to people.

Hemoglobin: A substance that gives the red blood cells their color, and carries oxygen around the body.

Hormone: A chemical messenger, made by an endocrine gland and sent around the body in the blood, which affects the way other parts of the body work.

Immune system: Your defense against germs. It consists of the various types of white cells found in the blood, including the cells that make antibodies and macrophage white cells that destroy germs.

Insulin: A hormone which helps to control the amount of sugar in the blood.

Lymph: The clear fluid which is carried in the lymphatic system and contains white cells important for fighting off germs.

Microbe: A term which describes all creatures too small for the eye to see without a microscope. The air, sea, and earth are full of these invisible creatures (mainly bacteria, fungi, viruses, and protozoa) most of which do not affect humans at all. Some are harmful to the body and are called "germs," whereas others, such as some of the bacteria in the intestines, are useful to the body.

Mineral salts: A mixture of metals, carbonates, nitrates, and phosphates which are needed in fairly small amounts to help the body to work. For example calcium, magnesium phosphates, and carbonates give bones and teeth their rigidity, and iron forms an important part of the oxygen-carrying hemoglobin.

Mitochondria: Small structures found in the cell that turn sugar into energy by a chemically controlled slow-burning process. They are the cell's "power stations".

Motor nerves: Nerves that carry messages from the brain and spinal cord to the muscles to make them work.

Mucus: A sticky liquid made by cells in areas such as the nose, breathing passages, and stomach. It is made of special proteins and water, and is useful as protection and lubrication.

Nervous system: The brain and nerves which control all fast action in the body, such as moving muscles and controlling organs.

Nucleus: The central part of the cell containing the genetic "instruction manual," packed into chromosomes. The nucleus directs the cell's activities.

Organ: A large collection of cells, often of several different types, which work together to do a particular job for the body: for example, the kidneys are organs which filter waste out of the blood and produce urine.

Oxygen: A gas needed by the whole body to convert food into useful energy. From the lungs it is carried around the body by the red cells in the blood.

Parasite: Any creature that lives off another one and harms it, without necessarily killing it. Some microbes are parasites, as are tapeworms.

Protein: A type of food that is needed for growth especially to build and repair muscle cells. Your body is also constructed mainly of proteins, and enzymes are a special type of protein.

Ribosomes: The "factories" in each cell that make the main parts of the cell proteins. They get their instructions from the nucleus.

Sensory nerves: Nerves that bring information to the brain from the sense organs: eyes, ears, nose, mouth, and skin.

Sperm: Sperm are special male cells that are made in the testes. If they break through the outer wall of the female egg they fertilize it to produce a new embryo.

Spleen: An organ that helps remove red blood cells when they die, and processes some white cells.

Starch: A carbohydrate food found in large quantities in bread, potatoes, rice, and pasta (all these foods also contain other substances, such as protein and vitamins).

Tissue: Groups of cells, usually of a similar type, which fit together to form an organ when combined with other tissues.

Vaccination: Putting a "mock" germ into the blood, so that the immune system can produce and develop defenses against it in case the real, harmful germ invades.

Veins: Tubes which carry blood to the heart. Only the vein which comes from the lungs has oxygenated blood in it; all other veins take blood back to the heart from tissues that have already used their oxygen.

Virus: A type of microbe that cannot live on its own because it feeds inside the cells of other creatures. Viral diseases include mumps, chicken pox, measles, and AIDS. Viruses are too small for the eye to see, even with an ordinary microscope.

Vitamins: Chemicals that you need regularly in small amounts to keep healthy (for example, vitamin C in fresh fruit and vegetables).

INDEX

A

accidents 34, 36
acne 43
acupuncture 46
adrenal glands 42
ageing 45, 52, 56-7
AIDS 54
alcohol 15, 56
allergies 26
alveoli 24
anger 42
ankles 30, 31
antibiotics 54, 55, 58, 60
antibodies 26-7, 60
appendix 15
arms 32, 33, 34
arteries 20-1, 23, 60
arthritis 55
artificial devices 33
asparagus 19
aspirin 15
axons 34, 35

B

babies 50-1, 56, 59
backbone see spine
bacteria 14, 15, 54, 58, 60
balance, sense of 39, 47
baldness 43
birth 50
bladder 9, 19
blindness 40
blood 7, 8, 58, 59, 60
    cells 10, 11, 20, 21, 26-7
    circulation 20-4
    digestive system 12, 13, 15
    in fetus 51
    groups 23
    HIV virus 54
    liver 18-19
    sugars 42
    wounds 45
blushing 23
bones 10, 11, 17, 28-31, 59
    breaks 31

brain 7, 8, 29, 34-47, 56, 59
    cells 10, 11
    embryo 50
    heart 22
breasts
    breast-feeding 42, 51
    cells 10
    puberty 57
breathing 8, 24-5, 36

C

calories 16
cancer 55
capillaries 20-1, 22, 60
carbon
    in body 59
    carbohydrates 16, 17, 59, 60
    carbon dioxide 24, 25, 60
cartilage 30, 55
cells 10-11, 20-1, 58, 60
    cancerous 55
    genes 51, 52
    immune system 26-7
    muscle 32-3
    nerve 34-5, 58
    reproductive 49
child development 56-7
clones 52
color vision 40-1
contraception 49
cramp 15
cystic fibrosis 52

D

dermis 45
diabetes 18
diaphragm 25
digestive system 7, 8, 12-16,
    18-19, 20, 36, 58
    see also food
diseases 17, 27, 52, 54-5, 59
DNA 11, 52-3, 60
Down's syndrome 52
dreams 38

drugs 54, 55
"dwarfism" 43

E

ears 39
egg 10, 11, 48-9, 50, 51, 60
elbows 30, 31, 33
"elephant man" 52
embryos 50-1, 60
endocrine systems 8, 42
enzymes, digestive 13, 14, 60
epidermis 44-5
epiglottis 13, 24
esophagus 12, 13, 14, 15
excretion 42, 60
eyes 36, 37, 38, 40-1
    color 53
    corneal grafts 33

F

faintness 23
fats 16, 17, 59, 60
fear 42, 54
fetus 50, 51, 60
fiber, dietary 17
fingers 34, 35
    fingerprints 44-5
    nails 50
food 16, 17, 46-7, 56
    see also digestive system
fractures 31
freckles 58

G

gall bladder 18-19
genes 11, 51, 52-3, 60
    lifespan 56-7
germs 26-7, 44, 54-5, 58, 61
glasses (spectacles) 41
glucagon 42
"goose pimples" 45

62